GEORGE WASHINGTON CARVER

GEORGE WASHINGTON CARVER
In His Own Words

SECOND EDITION

EDITED BY
GARY R. KREMER

UNIVERSITY OF MISSOURI PRESS

Columbia

ISBN: 978-0-8262-2139-1
Library of Congress Control Number: 2017951823

∞™ This paper meets the requirements of the
American National Standard for Permanence of Paper
for Printed Library Materials, Z39.48, 1984.

Typefaces: Abril, Caslon

Dedication, First Edition

For
Lorenzo J. Greene
My Mentor and Friend

Who has, and gives
Those qualities upon which friendship lives.
How much it means that I say this to you—
Without these friendships—life, what *cauchemar!*
—T. S. Eliot

Dedication, Second Edition

For Lisa A. Kremer and our children,
Randy, Sharon, and Becky,
and for our grandchildren,
Dustin, Brooke, and Bladen Kremer
Logan, Kaden, Halston, and Saylor Bax
Brennan Dazey
With All My Love

Contents

Acknowledgments

First Edition

WRITING THE "ACKNOWLEDGMENTS" section of a book is often one of the most difficult parts of completing a manuscript. For one thing, by the time an author gets to the stage where he is ready to commit his "thank yous" to paper, he owes so many debts of gratitude to so many people that he has at least the slightest twinge of conscience about claiming sole authorship to begin with. Secondly, he knows that if he mentions each person who has given him assistance the Acknowledgments section will rival in length the manuscript itself. Recognition of that fact, of course, leads to another anxiety: the author does not wish to offend anyone who has helped him by omitting his or her name.

The normal route out of this dilemma, the one that I also have chosen to take, is to state generally that this work could not have been completed without the assistance of a great many people whose help I truly appreciate. I want all of the persons who aided me in any way to know that I could not have produced this volume had it not been for their individual and collective support. In addition, I want to mention at least a few individuals who have given of themselves far more than an author has a right to expect of anyone.

The staff of the George Washington Carver National Monument has been extraordinarily helpful from the very beginning of this project. I am especially grateful to park Superintendent Gentry Davis and Rangers Larry Blake and Judy Bartzatt. I appreciate the fact that Monument officials have given me permission to reproduce Carver's letters.

Tuskegee Institute Archives officials have also been of great assistance, particularly Archivist Daniel T. Williams. I am also grateful to the Archives for permitting me to include Carver materials that they hold in their collection.

Likewise, I want to thank the publishers of *The Peanut Journal* for permission to reproduce two of Carver's essays that appeared in their magazine.

A sabbatical leave from Lincoln University allowed me to finish this manuscript, and I want to thank officials from that institution as well.

Linda Epstein typed and retyped substantial portions of the Carver manuscript in the early stages of its development, and Beth Wissmann performed the same invaluable function when the crunch came at the end. I want them to know how much I appreciate their patience and persistence.

Randall M. Miller of St. Joseph's College read an early draft of this work. His incisive and timely comments helped me to produce a far better manuscript than I had earlier attempted. Likewise, Linda O. McMurry of North Carolina State University, whose splendid book *George Washington Carver: Scientist and Symbol* helped me to organize and conceptualize this volume, has been of much help. She read and reread the manuscript and encouraged me to emphasize aspects of Carver's life that I had neglected. Linda Wyman, a Lincoln University colleague of mine, also labored over the entire manuscript with both wit and wisdom. I value the former as much as the latter.

Finally, I want to thank Lisa Dugan-Kremer and our children, Randy, Sharon, and Becky, for their patience and love. I know it can be difficult living with an author obsessed with preparing a book for press.

None of the persons mentioned above is responsible for any errors or shortcomings associated with this manuscript. That dubious distinction I claim as my own.

G.R.K.
Jefferson City, Missouri
August 1986

Acknowledgments

Second Edition

IT HAS BEEN difficult for me to come to grips with the fact that the first edition of this book was published three decades ago! I keep asking myself, "How can that be? Where has the time gone?"

I have also been surprised by the enduring interest in *George Washington Carver: In His Own Words*. The book has never gone out of print and continues to enjoy healthy sales.

Many times over the years I have thought about producing a new edition of this book. Primarily, I wanted to incorporate material I gathered from interviews I conducted with former colleagues, students, and friends of Carver shortly after the first edition came out. Those interviews form the basis of an entirely new chapter (Chapter 10). They were undertaken with the financial support of the Missouri Humanities Council, the Carver Birthplace Association, the Precious Moments Foundation, Inc., and the George Washington Carver National Monument. I appreciate the support of each of those institutions. Teena Vaughn and David Anderson of television station KOMU in Columbia, Missouri, and Rebecca Harriett of the Carver National Monument helped me to conduct the interviews, which I could not have done without them. The opportunity to talk with so many of Carver's contemporaries remains one of the greatest experiences of my life. Sadly, none of those persons I interviewed nearly three decades ago is still with us.

In 2015 I began to think seriously about preparing a new edition of *George Washington Carver: In His Own Words*, prompted in large part by the upcoming seventy-fifth anniversary of Carver's death (January 5, 2018), and turned to my friends at the University of Missouri Press. Clair Willcox, then associate director and editor in chief at the Press, encouraged me to take on the new edition, as did his successor as editor in chief, Andrew Davidson. Indeed, Andrew spent a significant amount of time reading and editing my work on this second edition. Mary Conley, Associate Acquisitions Editor at the Press, also helped bring this new work to publication, as did the director of the Press, David Rosenbaum. The marketing manager, Stephanie Williams, supported this new edition, just as she has other UM Press projects on which we have worked together.

I appreciate the help of my colleagues at the State Historical Society of Missouri. Senior Associate Director Gerald Hirsch scanned and OCR'd the original edition of the book, and Katherine Fallon corrected the OCR version and reformatted the original. Laura Jolley completed the index and Jeneva Pace and Melissa Wilkinson helped keep me on task in countless ways. I am also grateful for the support of the State Historical Society's Board of Trustees, especially President Bob Priddy and immediate past-president Stephen N. Limbaugh, Jr.

As was the case with the first edition of this book, the staff of the George Washington Carver National Monument has been extremely helpful. Park Ranger Curtis Gregory responded quickly and helpfully to every question I raised with him. He also apprised me of the Monument's latest acquisition of Carver materials: a collection of correspondence between Glenn Clark and George Washington Carver. Several of those letters are included in this new edition. I also appreciate Monument officials giving me permission to use materials, including photographs, from their collection.

Likewise, Tuskegee Institute archivist Dana Chandler helped me in innumerable ways. His deep knowledge of the Carver materials at Tuskegee was indispensable, and saved me a great deal of time. His assistant, Roderick Wheeler, was especially helpful in making photographs available for reproduction. I am grateful to Dana and the Tuskegee Institute Archives for permission to use letters, publications, and photographs from their collection.

I also appreciate the help provided me by the George Washington Carver Museum, part of the Tuskegee Institute National Historic Site, operated by the National Park Service. Park Ranger Shirley Baxter was especially helpful. Shirley was a student of mine at Lincoln University more than thirty years ago, when I first began this project.

Archivists Mark Schleer and Ithaca Bryant of the Ethnic Studies Center at Lincoln University's Inman E. Page Library also provided me kind assistance. Their facility houses a large collection of Carver materials on microfilm that made my task much easier by sparing me repeated trips to Tuskegee Institute. Bradley Kuenen, University Archivist at Iowa State University, also helped me.

Over the many years, several Carver scholars have been especially gracious in support of my effort to understand Carver. Linda McMurry Edwards and Peter Burchard have influenced my thinking on all things Carver through their important insights into Carver's life and times. Mark Hersey critiqued the first edition of this book, corrected some errors, pointed out omissions, and made important suggestions for improvement, most of which found their way into this second edition. No one has provided more help to me on this new edition than he, for which I am forever grateful.

Once again, I want to express appreciation to my family for their support and understanding of my passion for history. Lisa Kremer and our children, Randy, Sharon, and Becky, have always been supportive, as have sons-in-law Travis Bax and Jeffrey Dazey. The greatest blessing of my life has been to be the grandfather of Dustin, Brooke, and Bladen Kremer; Logan, Kaden, Halston, and Saylor Bax; and, Brennan Dazey. They are the Elite Eight, and they have brought more joy and inspiration to my life than I can ever adequately acknowledge.

Finally, I want to thank all of the readers of the first edition of *George Washington Carver: In His Own Words*. I hope that this new edition does not disappoint. If it does, there is no one to blame but me.

G. R. K.
Jefferson City, Missouri
May 2017

Editorial Policy

THE ORIGINALS OR copies of all of the letters contained in this book are housed either at the George Washington Carver National Monument in Diamond, Missouri, or at the Tuskegee Institute Archives in Tuskegee, Alabama. The letters in the Carver Monument collection tend to be of a more personal nature than those in the Carver Papers at Tuskegee. In addition, the Carver Monument letters are overwhelmingly letters written by Carver to his correspondents. By contrast, the Tuskegee collection features far more letters written to Carver than letters written by him.

The Carver materials are voluminous. The Tuskegee Archives alone contains more than one hundred and thirty Hollinger boxes of Carver's correspondence. Many Carver letters were destroyed in a disastrous fire at Tuskegee in 1947. Following the incident, Tuskegee officials launched an intensive effort to locate and acquire material pertaining to Carver's life. Consequently, the Tuskegee Archives contains numerous copies of letters, the originals of which are housed in other repositories. In each instance where I have used a copy of a letter, I have indicated where the original may be found.

This books is, necessarily, a selective collection of Carver's letters. A complete edition of Carver's correspondence would fill many volumes. I have tried to choose those letters that I thought would best reveal the complexity of Carver's personality and help to explain his attitudes and behavior.

I have *not* violated the integrity of Carver's inimitable writing style, nor have I altered his form in any way. Carver was often inattentive to the niceties of grammar, punctuation, and sentence structure. As he wrote to a friend on one occasion, "I am glad that you can read my letters. I was not aware that my spelling and grammar were even respectable."[1]

1. Carver to M. L. Ross, 7 April 1930; Tuskegee Institute Archives (TIA), George Washington Carver (GWC) Papers, reel 12, frame 0057.

Reading Carver's letters today, one gets the feeling that he wrote with such passion and intensity that slowing down to insure proper syntax or rules of agreement would have sapped his writing of its vigor. The ideas could not wait to get out. Indeed, he often wrote in incomplete sentences. Carver himself best summed up his attitude about the importance of substance over form when he told Dana Johnson, "I am more interested in the ideas expressed than the mechanics of writing . . . ."[2] To clean up Carver's writing would be an injustice to the reader and to Carver—he wrote the way he did because of the way he was.

I have in some cases deleted passages from the letters that seemed extraneous; all such deletions are indicated by ellipses.

In a few instances I have included writings by Carver that were not letters. For example, his testimony before the House Ways and Means Committee in the early 1920s, articles that appeared in *The Peanut Journal* and other contemporary journals and newspapers, and extracts from his Tuskegee bulletins, while not "correspondence," reveal much about the role he saw himself, and science generally, playing in American life.

2. Carver to Dana Johnson, 21 February 1931, cat. no. 1705, George Washington Carver National Monument (GWCNM).

Chronology

George Washington Carver: 1864-1943

[Compiled from information provided by staff of the George Washington Carver National Monument, Diamond, Missouri]

ca. 1864   Born, Diamond Grove, Missouri.

ca. 1876   Left Moses Carver farm to attend school in Neosho, Missouri.

1890   Enrolled at Simpson College to study piano and art.

1891   Transferred to Iowa Agricultural College (later Iowa State College of Agricultural and Mechanical Arts), Ames, Iowa.

1893   Paintings exhibited and received honorable mention at Chicago World's Fair.

1894   Bachelor of Agricultural Degree, Iowa State College of Agricultural and Mechanical Arts
Appointed member of faculty, Iowa State College.

1896   Master of Agriculture Degree, Iowa State College.
On October 8, went to Tuskegee Institute as Director of Agriculture at the invitation of Booker T. Washington.
Appointed Director of the Agricultural Experiment Station that had been authorized for Tuskegee by Alabama Legislature.

1906   On May 24, initiated Jesup Wagon with T. M. Campbell, Sr.

1916   Elected Fellow of the Royal Society for the Encouragement of Arts, London, England.

1921   Appearance, U.S. House of Representatives, Committee on Ways and Means, testifying in support of tariff on peanuts.

1923   Recipient, Spingarn Medal (NAACP) for Distinguished Service to Science.

1928   Honorary degree, Doctor of Science, Simpson College.

1935   Appointed Collaborator, Mycology and Plant Disease Survey, Bureau
            of Plant Industry, U.S. Department of Agriculture.

1937   On June 2, bronze bust of Carver unveiled on campus, a tribute from
            his friends throughout the nation for his forty years of creative research.

1938   Feature film, *Life of George Washington Carver,* made in Hollywood by
            the Pete Smith Specialty Company.
            Development of the George Washington Carver Museum by Board of
            Trustees of Tuskegee Institute.

1939   Recipient, Roosevelt Medal for Outstanding Contribution to Southern
            Agriculture.
            Honorary membership, American Inventors Society.

1940   George Washington Carver Foundation established at Tuskegee
            Institute.

1941   On March 11, George Washington Carver Museum dedicated at Tus-
            kegee Institute by Henry Ford, Sr.
            Recipient, Award of Merit by Variety Clubs of America.

1942   Honorary degree, Doctor of Science, Selma University, Alabama.
            Official marker for Carver Birthplace in Diamond, Missouri, autho-
            rized by the Governor of Missouri.

1943   On January 5, died, Tuskegee Institute, Alabama.

· ·

1943   His entire estate amounting to over sixty thousand dollars bequeathed
            to the George Washington Carver Foundation.
            78th Congress passed legislation H.R. 647, Public Law 148, creating
            the George Washington Carver National Monument, Diamond Grove,
            Mo. This legislation was sponsored by Rep. Dewey Short and Sen.
            Harry S. Truman.

1946   On January 5, 79th Congress-Joint Resolution, Public Law 290.
            Designated as George Washington Carver Day, issued by President
            Harry S. Truman of Missouri.

1947   Issuance of postage stamp in honor of George Washington Carver.
            George Washington Carver Museum fire (restored 1951).

1948   First day sale of the three-cent Carver commemorative Stamp.

1951   Fifty-cent piece coined to likeness of George Washington Carver and
            Booker T. Washington.

1952   Selected by *Popular Mechanics* magazine as one of fifty outstanding
            Americans and listed in their Fiftieth Anniversary Hall of Fame.

1956   Polaris Submarine *George Washington Carver* launched at Newport
        News, Virginia.
        Simpson College dedicated science building in memory of George
        Washington Carver.

1968   Iowa State College dedicated science building in memory of George
        Washington Carver.

1969   Elected to Agricultural Hall of Fame, Kansas City, Kansas.

1973   Elected, Hall of Fame for Great Americans.

1990   Elected, Inventor's Hall of Fame.

1994   Honorary Degree, Doctor of Humane Letters, Iowa State University.

1998   First day sale of thirty-two cent Carver Commemorative Stamp.

2005   Missouri Department of Agriculture Building named for George
        Washington Carver.

2015   Enshrined, Missouri Public Affairs Hall of Fame.

GEORGE WASHINGTON CARVER

ONE

Introduction

Carver—the Man and the Myth

No individual has any right to come into the world and go out of it without leaving behind him distinct and legitimate reasons for having passed through it.

Geo. W. Carver 25 May 1915

INTO THE TWENTY-FIRST century George Washington Carver remains a paradox of American history. He undoubtedly achieved one of his most sought-after goals: he made an indelible mark on the world he left behind. His name is as widely recognized as that of any other black American, past or present, with the possible exception of President Barack Obama, Martin Luther King, Jr., or sports figures such as LeBron James. He is a folk hero, our country's most well-known black success story—proof positive, we have all been told, that Horatio Alger is alive and well, even in twenty-first-century America. And yet, George Washington Carver is also one of the least understood of all our heroes. What manner of man was this person who, operating out of a remote southern black school, took white America as if by storm and rose to national and even international fame? How was he able to accomplish such a feat? The letters in this collection are presented in an attempt to answer those questions.

I knew little about George Washington Carver when I began a project of cataloging his correspondence for the National Park Service in 1982. I am of a generation that grew up after his death. As a child, I learned what most American schoolchildren were taught about him: he was born a slave, became a scientist, worked at Tuskegee Institute, and discovered countless ways to use peanuts. Carver was held out to me and my fellow students in the 1950s as an example of possibility and promise among the black race, a model for other blacks to emulate and an example that whites could point to whenever they wanted to prove that America was, indeed, the land of opportunity for all.

"In-depth" inquiries into the life of Carver during the 1940s and 1950s began and ended with a single book: Rackham Holt's *George Washington Carver: An American Biography*.[1] Holt's book reflected the view of Carver held by most Americans during the two decades after his death: it pictured him as a flawless, superhuman hero. Holt romanticized and mythologized her subject in an uncritical account of his rise from slavery to fame. The most important source she used in composing her chronicle of Carver's life was his own testimony of what he had accomplished. Holt visited him often in his Tuskegee office, interviewed him endlessly, and accepted as accurate the image of himself that he wished her to portray to the reading public. Carver read the several drafts of Holt's manuscript before it went to press and told her that it was "the most fascinating piece of writing that I have read. I started in and I confess I could not lay it down until I had finished it."[2]

As a college student in the 1960s, I learned a little more about Carver, but, frankly, his was not a life that many of my colleagues in black studies sought to understand. The very qualities that made him a hero to Americans of the 1940s and 1950s made him suspect among blacks and liberal whites in the 1960s and early 1970s. He was an "Uncle Tom," we said, easily dismissing him as a Booker T. bedfellow, and pronouncing him to be a subject unworthy of serious scholarly study, unless, of course, that scholarship was an instrument for ridicule. In 1972, Louis T. Harlan, the author of a two-volume biography of Booker T. Washington, barely took notice of Carver, referring to him as "an eccentric genius . . . noted for his quarrelsome nature, his loyalty to the school, and his deferential behavior to whites." Carver, according to Harlan, was undeniably useful to Washington, while "out-Bookering" the Tuskegee principal.[3] Criticism of Carver reached a crescendo with the work of a young historian named Barry MacKintosh with his essay, "George Washington Carver: The Making of a Myth."[4] Unlike Holt, MacKintosh found few redeeming qualities in Carver. Drawing upon such sources as a long-suppressed 1962 National Park Service report, which concluded that Carver's "discoveries" were greatly overrated, MacKintosh proceeded to flail away at the Carver image.[5]

Unfortunately for the readers of their respective works, neither Holt nor MacKintosh understood that Carver was not completely hero or myth. Instead, he was an extraordinarily complex man living in an extremely complicated society. The understanding of that reality was left to a historian of the 1980s, Linda O. McMurry, whose volume *George Washington Carver: Scientist and Symbol*[6] cleared up much of the confusion created by Holt and MacKintosh. McMurry avoided an unquestioning admiration of her subject on the one hand, while documenting and explaining, without condemnation, his shortcomings on the other. The conceptual framework provided by McMurry, undoubtedly influenced by the Civil Rights Movement of the 1960s, is the starting point for a more complete understanding of the man whom many dubbed "the Wizard of Tuskegee."

Since *George Washington Carver: In His Own Words* appeared in 1987, other scholars have taken up the challenge of describing and explaining Carver's place in twentieth-century American history. In 1998, Peter D. Burchard published a small book, *Carver: A Great Soul*, which emphasized Carver's spirituality and connection to nature. Burchard followed that up with A National Park Service-sponsored study titled *George Washington Carver: For His Time and Ours*, published in 2005.[7]

Two new books about Carver appeared in 2011. One was a book by this author, *George Washington Carver: A Biography*, published as a volume in the Greenwood Biographies Series by Greenwood/ABC-CLIO. The other was a creative monograph by a young environmental historian named Mark Hersey, who teaches at Mississippi State University. Hersey's book, *My Work Is That of Conservation: An Environmental Biography of George Washington Carver*, was published by the University of Georgia Press.[8] Hersey's book, influenced by the writings of Michael Pollan, among others, reflects the emergence and acceptance of environmental history as an important field of study. The great contribution of Hersey's work is that it transcends the old model of viewing Carver as an inventor and replaces it with a view of him as a conservationist. Indeed, Hersey argues, and correctly I think, that Carver's true genius and greatest contribution to history was as a pioneer conservationist.

More recently, biographer Christina Vella published a Carver biography in LSU Press's prestigious Southern Biography Series. Her *George Washington Carver: A Life* has been advertised as "the most thorough biography of George Washington Carver." Notwithstanding this claim, at least one reviewer of the biography concluded that despite Vella's "colorful and well-written biography," "George Washington Carver remains an enigma."[9]

My own reading of books about Carver, coupled with my exposure to the thousands of letters written by Carver housed at the George Washington Carver National Monument and Tuskegee Institute, continue to convince me that a collection of Carver letters can add detail to the portraits sketched by his biographers. Nowhere are the brilliance, self-doubt, religious fervor, and successes and failures of Carver more evident than in his own correspondence.

Although he never wrote in detail about it, Carver often referred to his origin as a slave.[10] The paradoxes and ironies surrounding his birth and early years seem appropriate for one whose life was enshrouded in such mystery. Carver was born a slave on a two-hundred-acre farm just outside the small town of Diamond in Newton County, Missouri. His master, Moses Carver, was a kindly Ohio-born transplant whose need for help around the farm overrode his opposition to the institution of slavery. George's mother, of course, lived on the Moses Carver property. His father belonged to a man who owned an adjoining farm.

George's father was killed in an accident before the birth of the future scientist. Subsequently, he and his mother were kidnapped by one of the many bands of bushwhackers who roamed western Missouri during the Civil War era. Moses Carver hired a neighbor to track down and rescue young George and his mother. The neighbor was only partially successful: he recovered the infant slave. The return of George Carver to his master's farm cost Moses Carver one of his finest horses. The mother disappeared, or at least seemed to disappear, from the pages of history, although there is inconclusive evidence that she may have reemerged after the war in a small northern Missouri town in which she spent the remainder of her life wondering about and looking for her son. If that Mary Carver was, in fact, George's mother, she died about the time her son went to Tuskegee. Carver often referred in his letters to the trauma that resulted from having been raised an orphan.[11]

In a 1989 interview, Dana Johnson, who as a young man had been a friend of Carver's, recalled visiting the latter in his room at Tuskegee.[12] He was struck by the presence of a small spinning wheel in the room, and Carver's high regard for it. "He told us that it had been his mother's and was one of the only things that he had of his mother's. If he touched it at all with his hand, it was with the greatest love and affection." Obviously, Carver felt the absence of his mother even in late life.

The death of his father and the disappearance of his mother meant that George and his older brother Jim would be raised by a former slave owner who had abolitionist sentiments. The first ten years of Carver's life are the sketchiest. He recalled very little of his childhood experiences. Late in his life, he offered this simple explanation for his faulty memory to Rackham Holt: "There are some things that an orphan child does not want to remember . . . ."[13]

The few recollections that Carver did have, combined with the remembrances of elderly Newton County residents, portray Carver's early years as a time when he was a frail, sickly child who, because of his poor health, spent much of his time assisting Susan Carver with domestic chores. While his brother Jim was out helping Moses Carver take care of the farm, George learned how to cook, mend, do laundry, embroider, and perform numerous similar tasks. Apparently George was still very young when he developed a fascination for plants, perhaps as a result of helping Susan Carver take care of the garden—another of the traditionally feminine tasks that his poor health and youth dictated he do. For the remainder of his life, George always found it easier to meet and talk with women than with men. In a late-life interview, Jessie Guzman, who worked with Carver at Tuskegee for more than two decades, recalled that he "had many women friends, mainly."[14]

One unfortunate by-product of Carver's early, continued, and extensive association with women was that it nurtured rumors that he was homosexual. The fact that he never married and that he had a decidedly feminine voice no doubt also provided fodder for the rumormongers. Despite Carver's close relationships with many young boys, there is no evidence that those relationships were anything other than platonic.[15]

On a more speculative note, Carver's fondness for things feminine may have influenced the way he practiced science. Students of the history of science agree that there are "feminine" and "masculine" approaches to scientific experimentation. The feminine approach is more intuitive and involves engaging in a dialogue with the subject being studied. Whether or not Carver's methods were affected by his stronger-than-usual ties to women, his statements about how the peanut told him of its potential uses, for example, fit the feminine model of the scientist fusing with the object of study and allowing that object to speak to the person who is studying it.[16]

Similarly, Elva Jackson Howell remembered a presentation that Carver made before a group of students at Virginia State College in 1928. She recalled how Carver placed a sweet potato on the lectern and began his talk by addressing the plant: "Sweet potato, sweet potato, what are you?"[17]

Susan Carver was not able to quench young George's thirst for knowledge. The same curiosity that prodded him to wander over Moses Carver's Diamond Grove farm in search of new flora and fauna led him also to leave the farm at about the age of eleven and travel eight miles to Neosho, where a school for blacks was conducted by a teacher named Stephen Frost.[18]

Carver arrived in Neosho too late at night to seek lodging with a friendly family, so he found a comfortable spot in a barn and settled in for the night. His choice of a sleeping spot was fortunate: first, the barn was practically next door to the school; second, it belonged to Andrew and Mariah Watkins, a childless black couple who took in the young waif and treated him as their own. George earned his keep by doing such chores as chopping wood, tending the garden, and helping the ever-busy "Aunt Mariah" with the weekly loads of laundry that she took in to help with the family's finances.

George's initial response to the opportunity for formal education was excited optimism, but his hope dimmed as he learned that Schoolmaster Foster knew little more than he did. He was happy enough with the Watkinses, but something was missing. So, like many a young person, before and since, he set out to find himself.

He hitched a ride with a family going west in the late 1870s and ended up in Olathe, Kansas.[19] For the next decade he traveled from one midwestern community to another, often using the domestic skills he had learned from Susan Carver and Mariah Watkins to survive. Doing laundry, for example, became his specialty. He even tried his hand at homesteading in Ness County, Kansas, in the mid-1880s. Like the other settlers in the area, all of whom were white, Carver built himself a sod hut and tried to eke out an existence on the recalcitrant Kansas prairie. Other Ness County folks quickly appreciated the fact that there was something special about this gentle black man who played an accordion for them at their dances, joined their local literary society, and showed a remarkable interest in and facility for painting. The correspondence contained in this collection reveals that Carver's Ness County memories remained precious to him throughout his life.

But there was something missing in Ness County as well. In the late 1880s wanderlust hit Carver again. This time his travels took him to Winterset, Iowa, where he encountered a white couple who profoundly influenced his life: Dr. and Mrs. John Milholland. Mrs. Milholland first noticed Carver singing at a church service one Sunday morning and, touched by his intensity and sincerity, sent her husband to fetch him home for Sunday dinner.

Conversations left the Milhollands deeply impressed: here, they knew, was a rare individual indeed. They quickly became convinced that the searching, sensitive mind of the future scientist needed to be nurtured and disciplined through formal education. They urged Carver to enroll as a student at nearby Simpson College, but he was reluctant. His only previous experience at entering college had ended disastrously. He had applied at Highland College in Kansas and been accepted, sight unseen. But when he had tried to register at the all-white school, the first official he encountered announced that there had been a mistake: Highland College had never admitted a Negro and had no intention of ever doing so. For a young man who had always considered whites to be his friends, that had to have been a bitter pill to swallow. He could not stand the thought of being rejected once more.

But the Milhollands persisted, prodding him to try again. They argued that his potential was so great that he owed it to himself. Finally, he gave in, moving to Indianola and enrolling in Simpson College in late 1889 or early 1890. He planned to study art, his first love. For the remainder of his life he was grateful to the Milhollands, often telling them he would never have enjoyed the benefits of higher education had it not been for them.

The correspondence between Carver and the Milhollands reveals the source of the "specialness" they and others saw in him. Carver's letters from his student days in the early 1890s are filled with references to an intimate relationship with God. He often wrote of spiritual obligations that needed to be carried out. God, he was convinced, had chosen him to perform wondrous tasks. Where Carver came by this deep sense of religion remains unclear. Back in Diamond, Missouri, he had had little formal religious training; Moses Carver had been a free thinker who distrusted organized religion and Bible-thumping preachers. Young George had gotten a good dose of regular Bible reading during his stay with Mariah Watkins, but his religious fervor seemed to stem more from a deep, personal mysticism—an almost pantheistic sense of identifying God with nature and communicating with Him through the forces of His creation.

This mysticism is one of Carver's most important traits and must be understood by anyone who would try to explain how and why he did what he did. He never separated the worlds of science and religion; he saw them as mutually acceptable and compatible tools for arriving at truth.

Throughout his life Carver had visions, which he took to be solemn directives for future action sent to him by God. His earliest recollection of a vision was as a child. He longed to have a pocketknife, but could not afford one. Then one night he had a dream that included a view of a knife lying in one of Moses Carver's fields. Upon awaking the next morning, he proceeded directly to the spot in the field that he had seen in his dream. There, sticking in a partially eaten watermelon, was a shiny pocketknife, the object of his greatest longing. Dreams, he quickly realized, were to be believed.[20]

Visions returned to him at Simpson College. He became convinced that God intended that he be a teacher of African Americans. Counseled by his art teacher, Etta Budd, who worried about his ability to make a living in the world of art, Carver decided to take a course in agriculture at the Iowa Agricultural College, where Miss Budd's father taught horticulture.

Encouraged by Miss Budd, and assisted by her father, Carver soon decided to abandon both Simpson College and the study of art, although he retained an interest in art and in painting throughout his life. He pursued an education in agriculture at Iowa Agricultural College, where he encountered a stellar faculty of "scientific agriculture" professors, including two future U.S. secretaries of agriculture, James Wilson and Henry C. Wallace. This was a practical decision: southern blacks could not paint their way out of poverty.

Carver made quite an impression on the Iowa Agricultural College faculty. His long-nurtured interest in plants had helped him to develop an ability to raise, cross-fertilize, and graft them with uncanny success. His professors were convinced that he had a promising future as a botanist and persuaded him to stay on as a graduate student after he finished his senior year. He was assigned to work as an assistant to Professor Louis H. Pammel, a noted mycologist, later described by Carver as "the one who helped and inspired me to do original work more than anyone else."[21] Under Pammel's tutelage, Carver refined his skills of identifying and treating plant diseases. As was the case with virtually all those who showed kindness toward him, Carver remained forever grateful for Pammel's help and encouragement, corresponding with him until Pammel's death and also maintaining a close friendship with the professor's wife and two daughters.

Carver was in the last year of his stay at Iowa Agricultural College when Booker T. Washington gave his famous Atlanta Exposition speech (1895). That speech was Washington's clearest expression of a long-held philosophy: that southern blacks needed to accommodate themselves to the reality of white control and win first their economic independence through vocational training and the ancient virtues of hard work and thrift. As a means to that end, Washington accepted a position as the first president of Tuskegee Institute in Alabama in 1881. By 1896 he had persuaded the board of trustees to establish an agricultural school. Carver, the only black man in the country who had graduate training in "scientific agriculture," was the logical choice for the Tuskegee leader, who wanted to keep his faculty all black.

So it was that late in 1896, George Washington Carver traveled to the struggling Tuskegee Institute, where he promised Booker T. Washington he would make grass grown green in the Alabama clay. In fact, while sitting in Washington's office he had another of his visions.[22] As the Tuskegee principal droned on about his aspirations for the school, and Carver's role in it, the young botanist gazed out an open window and looked into the future—both his own and that of the institution he had been called to serve. He saw a Tuskegee that was lush with green grass, a literal and figurative oasis in a barren desert—a place where burned-out southern blacks could come to be rejuvenated, a place of hope and promise. And always he would be there to show them the way—he would be their teacher. If they could not come to him, he would go to them; he would visit the poor and downtrodden in what he called "the lowlands of sorrow" and would help them learn to pull themselves up by their own bootstraps. He would combine the creativity of the artist with the rationality of the scientist to do what had never been done. Yes, Mr. Washington, he said, he was ready to begin.

But in reality he was not. In a sense, Carver could not have been more unprepared for what he found at Tuskegee. He expected to arrive there with something of a hero's welcome. Instead, he found that many persons at the Alabama school resented him. He was very dark-skinned, for example, much darker than most if not all of the other Tuskegee faculty and staff members. That simple physical reality made him immediately suspect among what E. Franklin Frazier would later call the "black bourgeoisie." He was coming from a northern state, having spent none of his adult life in the South; Tuskegeeans, who had lived all of their lives in one or the other of the former Confederate states, resented the fact that an outsider was coming to their school. They resented even more that he was going to be paid what they thought was an enormous salary, particularly for a man who had neither wife nor children to support. And, as if to add insult to injury, they heard that Carver had demanded two dormitory rooms in which to live: one for himself and one for his plant specimens, paintings, and other clutterings. This demand came at a time when other unmarried faculty members were bunking *two to a room*. Carver was also a notoriously shabby dresser. Indeed, his almost total inattention to his sartorial presence seemed a real slap in the face to his dress-conscious colleagues. Additionally, he was going to teach "scientific agriculture," which would put him in charge of people who were already convinced that they knew how to farm.

But one of Carver's biggest problems at Tuskegee was that he simply was not used to living with and working around other blacks. Most of the people who had touched his life in a significant way, with the exception of his light-skinned brother Jim and the Watkinses, had been white. In fact, at least two sources have reported that it was on an early trip to Neosho, when George was approximately six years old, that he came in contact for the first time with truly dark blacks, an encounter that caused the young traveler to flee in terror.[23]

Whites had always nurtured Carver's image of himself as unique. They tended to see him, in the nineteenth-century phrase reserved for but a few blacks, as a "representative colored man"—a model that they could point to as an example of what the race could and should become. They did not fear him; he seemed to them to be so different from the black masses that the difference was one of kind rather than degree. The more praise they heaped upon him, the more he sought it. Later in his life, he liked to point out to his correspondents that he was frequently invited to speak to white groups. He relished the favorable comments about him made by whites—the more powerful they were, the better he liked their attention. He was particularly proud of his friendship with Henry Ford, for example.

But Carver's solicitousness of whites offended his Tuskegee colleagues. And the more they criticized him, the more he curried white favor. It was a vicious circle that he never learned to escape.

So, all things considered, the start at Tuskegee was bound to be slow. Carver expected that as director of the newly established Agricultural Experiment Station he would spend most of his time doing research. Washington did want him to do research, but he also wanted Carver to manage the school's two farms, teach a full load of classes, and serve on numerous committees and on the institute's executive council. Washington even expected him to oversee the proper functioning of the school's water closets and other sanitary facilities. And all of this was to be done when he was not busy with the nearly full-time administrative chores that went with heading the new Agricultural Department.

The pragmatic, goal-oriented, no-nonsense Washington quickly became impatient with the unorganized, administratively inattentive, idealistic Carver. As Linda McMurry has written, Washington was a realist and Carver a dreamer.[24] Carver wanted the freedom to piddle and putter in his lab, experimenting with this and that plant, or, when the mood struck him, picking up a brush and his paints and playing the artist for the afternoon.

Not surprisingly, a great deal of friction emerged between Carver and Washington, the principal often complaining that a report was late, that Carver was not spending enough time at his administrative chores, or that he was managing poorly the people who worked under him. Carver's response, invariably, was that too much was expected of him and that his facilities and resources were totally inadequate.

The bitterest feud in which Carver became embroiled at Tuskegee was with George R. Bridgeforth, a subordinate who sought to take his place. Bridgeforth, a native of Alabama who was educated at the Massachusetts Agricultural College in Amherst, was more than a decade younger than Carver. Like Carver, an outsider educated at a northern white school, Bridgeforth might have been a natural Carver ally. Instead, he openly criticized his boss to Washington and wrote acerbic letters to Carver that included comments such as the following:

> As I have told you in person, I feel very bad over the affair and cannot yet see how you can use me in this way after I have done so much for you and worked so hard to get these improvements made. I have in the last few weeks been greatly embarrassed by you telling me to do things and then could not back me up after I had done as you commanded.[25]

Carver did not expect anyone to talk to him like that, least of all an underling. To make matters worse, he found that Washington often took Bridgeforth's side and echoed the younger man's criticism. That was almost more than Carver could take, and instead of swallowing his pride and accepting Washington's suggestions, which, the principal acknowledged were "but a polite way of giving orders," Carver continuously challenged him.[26]

Years later, in the 1970s, after both Washington and Carver were dead, many of their contemporaries had forgotten much of the trouble on the Tuskegee campus in the early twentieth century. Some even remembered the principal and the scientist as being universally admired by all who worked with them. But other of Carver's former coworkers remembered things differently. In 1974, Irving Menafee, who had worked with both Bridgeforth and Carver, was still of the opinion that "He [Carver] ain't learned nobody nothing. I never did think too much of Dr. Carver cause all the things he put out mostly benefitted white people." Carver, Menafee contended, "Never did go out much. . . . And I tell you something else between me and you. Dr. Carver didn't learn no colored people nothing. He didn't trust nobody."[27]

Hattie West Kelley, a teacher of music at Tuskegee, recalled that Carver was generally inaccessible both to faculty and students: "Unless you pushed your way into him, he would not see you."[28] Mrs. Elizabeth Ray Benson, who grew up on the Tuskegee campus, recalled that Carver was often oblivious to those around him: "You would meet him out there; he might not even see you. . . ."[29] Jessie Guzman, who came to Tuskegee in 1923 and worked in the Department of Records and Research, remembered, "When I first came, nobody thought too much of Carver. . . . they didn't take Dr. Carver very seriously . . . ."[30]

Early in his stay at Tuskegee, Carver focused his efforts on the classroom and on teaching rural southern blacks, especially those in Macon County, how to improve their lives through better farming techniques. He experimented with paints that could be made from Alabama clay because he knew that southern yeomen could not afford commercial varieties. He tested organic fertilizers because he realized that man-made revitalizers of the soil were also beyond the ability of most southerners to purchase. And, most important, he continually searched for new foodstuffs that would provide variety and supplement for the dirt farmer's diet. The peanut, of course, fell into this category, and it was the peanut that made George Washington Carver famous.

Carver had sought to disseminate information about his discoveries with the peanut, and a host of nature's other gleanings, to Macon County residents early in his Tuskegee career. Those efforts culminated in May 1906 when he sent out into the countryside the "Jesup Wagon" loaded with products that came out of his laboratory. Over the next decade, his reputation as a chemist and an agriculturalist grew, earning him an election to a Fellowship in the

English Royal Society for the Encouragement of Arts in 1916. As his fame grew, his work in the classroom and on behalf of "the man farthest down" diminished.

The most important event that caused Carver to move into the forefront of American consciousness was his appearance in 1921 before the House Ways and Means Committee as an expert witness on a pending bill that proposed to place a tariff on the peanut. Carver began his presentation before the committee by demonstrating the variety of foods that could be made from the lowly legume. Initially he operated under a ten-minute time limit, but he so captivated his audience that the time limit was repeatedly extended, until the committee chairman announced that Carver could take as long as he needed. Carver spoke for nearly an hour, and at the end of that period, as Linda Mc-Murry has written, he "had won a tariff for the peanut industry and national fame for himself."[31]

Successful testimony before a congressional committee brought Carver attention and acclaim that could never have been achieved at Tuskegee, no matter how many miles the Jesup Wagon traveled or how many new products he discovered. George Washington Carver had become a national folk hero.

For the remainder of his life, the Carver mystique continued to grow. He gave Americans what they seemed to need most: tangible evidence that anyone, even a black man born into slavery, could achieve success. He was, at least on the surface, what all Americans wanted to believe they were: self-deprecating servants to the common good. He was a black "Everyman," as every man wanted and claimed to be. Thomas Edison sought his services and offered, according to reports leaked intentionally by Carver to the press, a six-figure salary if only the black genius would come to Menlo Park. But Carver, in true hero fashion, declined the offer so that he could remain in the service of "his people."[32]

The Carver myth continued to build. Stories surfaced about how this unassuming man was so unconcerned with material gain that his secretary periodically searched his desk drawer for the half-dozen or so monthly pay checks that he was too busy, and too unconcerned, to cash.[33] And when the Great Depression hit with all its force in the mid-1930s, Carver added to his own mystique by nonchalantly writing to friends that the failure of the Tuskegee Bank had cost himself, the bank's largest depositor, more than $30,000 in personal savings. He was disturbed, he wrote, because the bank's failure made it impossible for him to give money to needy people he ran into on the streets of Tuskegee.[34] When a friend wrote to ask him if he was worried about the situation, he replied, "No, I am not worrying about it. Just disgusted, with the hope that I will get at least some of it by and by."[35]

The Depression, in a sense, was a godsend for Carver's fame. In an age of real scarcity, an ever-increasing number of people turned to him as an expert on conservation and how to make do. Even the Russians solicited his help with their agricultural program in the lean thirties.[36]

The fame, and the preferential treatment that came with it, only exacerbated tensions between Carver and his Tuskegee critics. There were those among the group with which he worked daily who persisted in resenting him, even until his death. In fact, for some the resentment grew stronger in proportion to the attention he received.

Not everyone at Tuskegee disliked Carver, of course. Those who knew him best tended to like him best, such as Harry Abbott, a Tuskegee colleague who became a good friend and who never found fault with him. Abbott served as the famous scientist's traveling secretary in the early 1930s, when Carver was making extensive speaking tours throughout the South. Subsequently, Abbott left Tuskegee and went to work in Chicago, and he and Carver began a correspondence that lasted until the latter's death. That correspondence gives insight into Carver's relationships with his friends and his feelings about Tuskegee.

Abbott disapproved of the way Carver was treated at Tuskegee. In October 1964, when Abbott turned his correspondence from Carver over to George Washington Carver National Monument officials, he included a cover letter that carried this thought:

> I do not intend to send any of the things I have to Tuskegee. I resent the way the younger (post-Moton) crowd treated Carver. He sensed it and that was one of the things he liked about me. I always felt very humble both with and away from him and I know how much he appreciated it.[37]

Carver was intense in his friendships. He gave loyalty, attention, even devotion, to his friends; he expected no less in return. He invited Abbott to worry over him, for example, rarely missing an opportunity to comment on his poor health in the late 1930s. In September 1937, he wrote to tell Abbott that it was good to hear from him, inasmuch as Carver was quite weak and ill. He confided that he had been battling an undefined malady for several months and that "very, very few people realized how far down I was, as I kept going when many people would have been in the hospital."[38] In January 1938 he wrote that he was still very weak and that his heart was "giving me warning every little while to the effect that it may just get tired and stop for a long rest."[39] Carver was planning a trip to Minneapolis in the spring of 1938 that would take him through Chicago and allow a visit with Abbott. But on 30 March

1938, he wrote, "Yesterday I had a complete collapse which makes it absolutely imperative that I cancel the trip to Minneapolis, much to my regret . . . ."[40] Just three days later, however, after this "complete collapse," Carver wrote to tell Abbott that the trip was on again: "I plan to leave Tuskegee by way of Chicago for St. Paul, Minnesota, Monday evening. We shall hope to call you up when we reach Chicago."[41] Carver apparently did not make the trip; letters of 26 and 28 April make no mention of it. The 28 April letter does contain the following comment: "Physically I have been in much worse condition than most people imagine. Because they saw me moving around a little daily, they said 'Oh, he is all right,' and they gave out that information."[42]

Soon after writing that letter, Carver was confined to the hospital for an extended rest. Although a hospital spokesman wrote to Abbott to assure him that Carver's condition was not "alarming," some months later Carver wrote to Abbott, "I am not complaining by any means as it has leaked out now very definitely that I was not expected to come out of the hospital alive, and it has surprised everyone, even the physicians . . . ."[43]

The good thing about his illness, as Carver saw it, was that it caused Tuskegee officials to notice him. "In fact," he declared, "people are beginning to feel just a little bit alarmed and are giving me the protection now that I should have had a long time ago. They are keeping people from crowding me and worrying me nearly to death."[44]

At times, Carver became impatient with the frequency with which Abbott wrote. On 23 August 1938, he wrote a two-line letter to Abbott that said simply, "Please tell me what has happened to you. I have not heard from you in so long that I am uneasy."[45] Again in October, Carver had not heard from Abbott for some time, and he wrote, "I have not heard from you for so long, I am anxious to know how you are getting along. Don't take time to write a long letter, but just let me know how you are."[46] That letter prompted an Abbott response that Carver received on 14 October. Abbott wrote again, less than two weeks later. But, as far as Carver was concerned, twelve days between letters was too long. He began a letter on 27 October with the following gentle chastisement:

> Your most interesting letter reached me yesterday and made me unusually happy as I have been expecting a letter for some time. I know that you keep unusually busy, but when I don't get a letter once every so often, I naturally get lonesome, as I do not know just what has happened to you, and naturally, I don't want anything to happen to you. I am endeavoring to adjust myself to your being away as best I can, but it is a hard job.[47]

In February 1940, he wrote even more caustically:

Well you have relieved a most severe tension. I was almost frantic when I could not hear from you. I had all sorts of misgivings. In fact, I dreamed one night that your mother was dead and that was the reason you did not write. However, please don't let this happen any more. Just drop me a card saying that all is well if you haven't time for more.[48]

Carver even found something to complain about when he received a lengthy letter from Abbott in November 1941: "I would like to make the suggestion that you don't wait so long and write such a long letter, but divide them into just about half and send them oftener as it does seem such a long wait before I hear from you."[49]

Among the most revealing correspondence in this collection are the letters to two of Carver's "boys," Jimmie Hardwick and Dana Johnson. By the late teens and early twenties, Carver had gained attention for his work throughout the South and was often invited to speak to white student audiences. This was especially true of functions associated with either the Atlanta-based Commission on Interracial Cooperation (CIC) or the Young Men's Christian Association (YMCA), groups committed to improving interracial understanding. When Carver spoke to groups of young people, he often searched the audience for faces that bespoke intense interest in his message. One such face was that of Jimmie Hardwick, whom Carver saw in a group of students he was addressing at a YMCA summer school in Blue Ridge, North Carolina, in 1922. Hardwick, the descendant of slaveowners, had been the captain of the Virginia Polytechnic football team and was searching for a way to be of Christian service. He walked up to Carver after the scientist's presentation one evening and told him that he would like to talk further with him. Carver spoke with a suddenness and a bluntness that frightened the young Hardwick: "Of course! I'd like you for one of my boys."

Two days passed before young Jimmie Hardwick dared approach Carver again. Then one night, after he had retired to his room, Carver heard a knock at his door: it was the young Hardwick, come to talk. He began slowly, with chitchat about Tuskegee and what Booker T. Washington had been like. Finally, Hardwick asked the question he had come to ask: "Professor Carver, just what did you mean when you said you wanted me to be one of your boys?" Carver's response was direct and honest: "In my work, "he said, "I meet many young people who are seeking truth. God has given me some knowledge. When they will let me, I try to pass it on to my boys." Hardwick's response was equally forthright and honest: "I'd like to be one of your boys, Professor Carver, if you will have me."[50]

Hardwick did become one of Carver's boys, of course, and this collection contains numerous letters from Carver to Hardwick, written between 1923 and 1933. That Hardwick idolized Carver is clear enough; but it is also obvious that Carver derived as much comfort and pleasure from the relationship as did the young Virginian. It is impossible to read the Hardwick letters and remain unconvinced of Carver's capacity for intense friendships. It is even more striking that an elderly world-famous scientist could have such a friendship with a young, unknown student, barely a third his age. Still, the Hardwick letters, like Carver himself, are a curious mixture of opposites. The language of the letters is intimate and even affected. Yet each letter addresses the young disciple with the formal salutation of "Mr." Hardwick. Carver, no doubt influenced by contemporary racial etiquette, could not bring himself to greet one of his closest friends by a first name.

A deep friendship is also evident in the correspondence between Carver and Dana Johnson. The Johnson letters in this volume represent a collection that became available to scholars during the early 1980s. Johnson, a retired chemist who lived in Newport Beach, California, until his death in 2006, donated twenty-eight letters to the Carver National Monument in 1982.

Johnson and his brother Cecil had both been Carver's "boys." Cecil met Carver first as a summer (1929) employee of the Tom Huston Peanut Company, a firm to which Carver served as a consultant. Carver invited Cecil to visit him at Tuskegee, and Cecil brought Dana along for the forty-two-mile trip from Columbus, Georgia. Both Cecil and Dana were students at Georgia Tech.

Dana Johnson met Carver on that visit on 1 January 1930. He and Cecil walked into a room in which an artist named Isabel Schultz was working on a bas-relief model of Carver. Dana Johnson was intrigued immediately by the sculpture and observed the work intently. Carver, noting Johnson's interest, asked him what he thought, to which Johnson replied that the nose was not quite straight enough. The aging scientist was captivated by the young man's honesty, intensity, and attention to detail. A close friendship that would last the remainder of Carver's life had begun.[51]

Carver continuously nurtured Dana Johnson's artistic aspirations. Letters to Dana urged him to care for his physical, mental, and spiritual health, to avoid worry, to trust in God, and to develop the artistic abilities that God had given him. Rarely did Carver miss an opportunity to heap praise upon both Cecil and Dana and to tell them how precious and special they were to him.

Comments such as the following are commonplace: "You two dear fellows are my ideal of young men from every angle"; "I love those creative minds that are destined to make a mark in the world."[52]

At times, Carver went to great lengths to analyze and critique pieces of artwork sent to him for evaluation by Dana. He offered advice on technique, of course, but he also commented on the feelings evoked by certain pieces and discussed with Dana the emotions and souls of poets, writers, and artists.

One topic that surfaces often in the correspondence to Dana Johnson is that of Carver's treatment of paralysis victims with peanut oil. Carver's concern for health and his belief in the curative powers of the things of nature had been with him at least since his late-childhood days in Neosho, Missouri. Mariah Watkins had been the community midwife and country doctor whose knowledge of "simples" and other herbal remedies were relied upon by Newton County residents. As a student at Ames, Iowa, Carver, who remained an avid football fan all his life, became masseur to the Iowa Agricultural College football team, an experience that increased his belief in the beneficent results to be obtained from a massage.

Early in his career at Tuskegee, Carver began treating his friends to massages with peanut oil. Simultaneously, polio was becoming more and more dreaded as each summer it raged through the country, attacking the young especially. A vaccination against the crippler of thousands would not be discovered until 1955. But if the disease could not be stopped in Carver's lifetime, he, at least, became convinced that he could treat its effects and restore life to withered limbs.

The foundation for Carver's belief in the power of the peanut was a side effect he observed among people who used his peanut oil-based facial creams. Women who employed these beauty aids complained to him that their faces looked fat, the result of the cream soaking into the skin and causing it to expand. Carver became convinced that the peanut oil not only saturated the skin and flesh but actually entered the blood stream if massaged onto the body properly.

Carver was further convinced that the miracle cure he had discovered was not the result of an accident. Persuaded that the guiding hand of Providence always led him in certain directions, he saw the massage as simply another way in which God was using him as an instrument for human progress. His first self-proclaimed success with the peanut oil massages came in the mid-1920s when the sickly, emaciated son of a prominent white Tuskegee family began to visit him three times a week for treatments. Within thirty days the youth had gained thirty-one pounds and was well on his way to recovery.

By decade's end, Carver had begun experimenting with the treatment on a few polio victims. He hinted at his "discovery" in speeches and then, on 30 December 1933, the Associated Press carried a story about his tentative successes.[53] Subsequently, Tuskegee became a Lourdes-like sanctuary for paralytic pilgrims.

Whether or not Carver's patients benefited from his peanut oil massages remains open to debate, although he frequently cited individual cases in which he provided what he thought was conclusive proof that they had. And whatever success he achieved with persons who sought his help, Carver was quick to point out, was due to God's work through him.

The letters Carver wrote to his many correspondents over the course of his life reveal much about the complex personality of the man. He could make petty and unreasonable requests of his friends. He was a difficult person to work with, and he had a barely controllable ego. But he also possessed a sincerity and a capacity for affection that captivated many of the people he met. He could be simultaneously selfish and selfless. He believed in his own creative brilliance, but he believed just as strongly that the source of that brilliance was a force over which he had no control.

Perhaps the complexity and richness of the Carver personality have so eluded us because he defies easy categorization. He does not fit comfortably into the molds we generally use for the purposes of description. Was he a scientist? Of course, but he also relied on very unscientific means of obtaining knowledge. Religious inspiration was as much a part of his search for truth as was experimentation. Was he successful? Yes, but he refused to measure success in the terms used by those around him. "It is not the style of clothes one wears," he once said, "neither the kind of automobiles one drives, nor the amount of money one has in the bank that counts. These mean nothing. It is simply service that measures success."[54]

George Washington Carver was a deeply religious man who treasured the world of nature and saw himself as a vehicle by which the secrets of nature could be understood and harnessed for the good of mankind. That was his mission in life, and his reward for performing this mission was the simple knowledge that he was performing well God's will.

At least that was the way Carver wanted it to be. But he needed almost constant reinforcement from the world of men for the job he was doing for God. He was insecure and full of self-doubt. His orphan origins and white American racism had arguably laid the foundation for that insecurity; the presence and persistence of hostility toward him at Tuskegee built upon the foundation.

Being accepted by whites was not only important to Carver, but it validated his very existence. He never felt truly comfortable with "his people." Instead, he was more like a missionary to them, and, like so many missionaries, he expected to be treated with gratitude. Many of his Tuskegee colleagues, were unimpressed. At best they saw him as an intruder and a prima donna; at worst, they thought him a traitor. Not surprisingly, he turned more and more to whites for approval; that thirst for praise drove him to work hard, speak often, and travel much. By the 1920s and 1930s, Carver eagerly accepted the plaudits of a diverse swath of Americans, who celebrated him for wildly divergent reasons, ranging from his agricultural "discoveries" to his treatment of the consequences of infantile paralysis.

And what did Americans get for showering a black scientist with praise? African Americans found a source of pride, an indication of the latent abilities of a race that could not be permanently kept down by political and economic oppression. White Americans, on the other hand, received justification for the status quo and confirmation that the country was, in fact, on the right track. In an era of great transition, when a war that everyone said would not happen occurred and when the land of plenty went broke, few things seemed certain anymore. Americans were in search of something or someone that tied them to what they perceived as their traditional roots. They found that someone in the person of George Washington Carver. The land of opportunity for all— always acclaimed but rarely realized—elevated a black ex-slave to the status of hero. His life was proof that America was still on the right track.

There was no risk involved in raising Carver to the level of hero: he was inoffensive and harmless. Though he was often unhappy with the status quo, he seldom directly challenged it. He was content to accept racial strictures, convinced that time, black accomplishments, and a Christian spirit would eventually change all.

Americans could scarcely wait for Carver to die to make his birthplace into a national monument, the first such honor they bestowed on a black person. The old scientist would have relished the symbolism involved in such a tribute. It meant that he had been right all along. He had left behind him "distinct and legitimate reasons" for having passed through the world. He really had been special.

TWO

Self-Portraits

Carver's Self-Image over Time

> It is impossible for me to tell you how astonished I was yesterday to get a copy of the November issue of Junior Red Cross journal bearing the illuminating article on "The Sage of Tuskegee." As I read it a number of times I became so interested that I forgot that it was about myself. In fact, the story is so beautifully and interestingly written that it is hard for me to believe that it is about me. I lose that part of it.
>
> Geo. W. Carver 29 October 1936

GEORGE WASHINGTON CARVER assiduously cultivated an image of himself as a poor defenseless orphan who, from his very earliest days, had been blessed by God with an inexplicable thirst for knowledge and a penchant for scientific discovery. He was special, but he was also fragile. Rarely was he content to allow his accomplishments to stand on their own merit; rather, he insisted on measuring them against a whole chain of adverse circumstances he had had to fight against. He seemed to want to say, to no one in particular, or rather to anyone who would listen, "You are right to be impressed with what I have accomplished, but if you only knew the odds against which I struggled, you would be astounded."

And impress people he did. As early as 1897, his first full year at Tuskegee, two of his ex-teachers at Ames wrote to him and asked him to pen a brief sketch of his life. Reading that autobiographical statement today, one marvels at the powerful self-image portrayed by someone whose professional career quite literally lay ahead of him. Confidence, even benign arrogance, permeates the piece. But there is also a sense of exaggeration bespeaking a deep insecurity. How, for example, could a sickly boy whose existence was nothing more than "a constant warfare between life and death to see who would gain the mastery" find the energy to spend day after day "in the woods alone in order to collect my floral beautis"? And there is a desire to tell the story while feigning too much modesty to want to do so. One way of looking at Carver's self-image is to assess his autobiographical statements and his reaction to the many efforts made by a variety of people to tell his life story.

Here, then, is Carver's earliest written reflection on his early life.

1897 or thereabouts[1]

As nearly as I can trace my history I was about 2 weeks old when the war closed. My parents were both slaves. Father was killed shortly after my birth while hauling wood to town on an ox wagon.

I had 3 sisters and one brother. Two sisters and my brother I know to be dead only as history tells me. Yet I do not doubt it as they are buried in the family burying ground.

My sister mother and myself were ku *klucked*, and sold in *Arkansaw* and there are now so many conflictinng reports concerning them I dare not say if they are dead or alive. Mr. Carver the jentleman who owned my mother sent a man for us, but only I was brought back, nearly dead with whooping cough with the report that mother & sister was dead, although some sauy they saw them afterwards going north with the soldiers.

My home was near Neosho Newton Co Missouri where I remained until I was about 9 years old my body was very feble and it was a constant warfare between life and death to see who would gain the mastery - - - -

From a child I had an inordinate desire for knowledge, and especially music, painting, flowers, and the sciences Algebra being one of my favorite studies

Day after day I spent in the woods alone in order to collect my floral beautis and put them in my little garden I had hidden in brush not far from the house, as it was considered foolishness in that neighborhood to waste time on flowers.

And many are the tears I have shed because I would break the roots or flower of off some of my pets while removing them from the ground, and strange to say all sorts of vegetation succeed to thrive under my touch until I was styled the plant doctor, and plants from all over the country would be brought to me for treatment. At this time I had never heard of botany and could scerly read.

Rocks had an equal facination for me and many are the basketsful that I have been compelled to remove from the outside chimney corner of that old log house, with the injunction to throw them down hill. I obeyed but picked up the choicest ones and hid them in another place, and some how that same chimney corner would, in a few days, or weeks be running over gain to suffer the same fate I have some of the specimens in my collection now and consider them the choicest of the lot. Mr. and Mrs. Carver were very kind to me and I thank them so much for my home training. They encouraged me to secure knowledge helping me all they could, but this was quite limited. As we lived in the country no colored schools were available so I was permitted to go 8 miles to a school at town (Neosho). This simply sharpened my apetite for more knowledge. I managed to secure all of my meager wardrobe from home and when they heard from me I was cooking for a wealthy family in Ft Scott Kans. for my board, cloths and school privileges.

Of course they were indignant and sent for me to come home at once, to die, as the family doctor had told them I would never live to see 21 years of age. I trusted to God and pressed on (I had been a Christian since about 8 years old.) Sunschine and shadow were profusely intermingled such as naturally befall a defenceless orphan by those who wish to prey upon them

My health began improving and I remained here for two years, From here to Olutha Kans. to school, From there to Paola Normal School, from there to Minneapolis Kans. where I remained in school about 7 years finishing the high school, and in addition some Latin and Greek. From here to Kans. City entered a business college of short hand and type-writing. I was here to have a position in the Union telegraph Office as stenographer & typewriter, but the thirst for knowledge gained the mastery and I sought to enter Highland College at Highland Kans. was refused on account of my color.

I went from here to the Western part of Kans. where I saw the subject of my famous Yucca & cactus painting that went to the Worlds Fair. I drifted from here to Winterset Iowa, began as head cook in a large hotel. Many thanks here for the acquaintance of Mr. and Mrs. Dr. Milholland, who insisted upon me going to an Art School, and choose Simpson College for me.

The opening of school found me at Simpson attempting to run a laundry for my support and batching to economize. For quite one month I lived on prayer beef suet and corn meal. Modesty prevented me telling my condition to strangers.

The news soon spread that I did laundry work and realy needed it, so from that time on favors not only rained but poured upon me. I cannot speak too highly of the faculty, students and in fact the town generaly, they all seemed to take pride in seeing if he or she might not do more for me than some one else.

But I wish to especially mention the names of Miss Etta M. Budd (my art teacher Mrs. W. A. Liston & family, and Rev. A. D. Field & family. Aside from their substantial help at Simpson, were the means of my attendance at Ames.

Please fix this to suit)

I think you know my career at Ames and will fix it better than I. I will simply mention a few things

I received the prize offered for the best herbarium in cryptogamy. I would like to have said more about you Mrs. Liston and Miss Budd but I feared you would not put it in about yourself, and I did not want one without all.

I received a letter from Mrs. Liston and she gave me an idea that it was not to be a book or anything of the kind this is only a fragmentary list. I knit Chrochet, and made all my hose mittens etc. while I was in school.

If this is not sufficient please let me know, and if it ever comes out in print I would like to see it.

God bless you all.

Geo. W. Carver

Twenty-five years after Carver wrote the above account, he penned another sketch of his life. There are differences in detail in the stories: in 1922, for example, Carver recalled the name of the neighbor who owned his father; he knew the identity of the man sent to retrieve him from his "Ku Klux" captors and also recalled that his ransom was "a very fine race horse and some money"; he remembered that his only book as a child was a Webster's Elementary Spelling Book; and he erroneously identified the college that discriminated against him as a school in Iowa, rather than Kansas. He also had come, by 1922, to regard Dr. and Mrs. John Milholland, of Winterset, Iowa, as "my warmest and most helpful friends."

But the overall theme of the two stories is remarkably similar: a sickly slave boy, orphaned by the disappearance of his mother, he grew up in the home of benevolent whites and displayed a savant's thirst for knowledge. His longing for formal education led him on an odyssey through three states, with white philanthropists and well-wishers assisting him every step of the way, despite an overpowering sense of modesty that precluded him from ever asking directly for help: "I would never allow anyone to give me money no difference how badly I needed it."

One of the many interesting things about the 1922 autobiography is that it contains but two sentences regarding Carver's career at Tuskegee, despite the fact that he had, by that time, been at the Alabama school for more than a quarter of a century.

A Brief Sketch of My Life[2]

I was born in Diamond Grove, Mo., about the close of the great Civil War, in a little one-roomed log shanty, on the home of Mr. Moses Carver, a German[3] by birth and the owner of my mother, my father being the property of Mr. Grant, who owned the adjoining plantation. I am told that my father was killed while hauling wood with an ox team. In some way he fell from the load, under the wagon, both wheels passing over him.

At the close of the war the Ku Klux Klan was at its height in that section of Missouri. My mother was stolen with myself, a wee babe in her arms. My brother James was grabbed and spirited away to the woods by Mr. Carver. He tried to get me, but could not. They carried my mother and myself down into Arkansas, and sold my mother. At that time I was nearly dead with the whooping cough that I had caught on the way. I was so very frail and sick that they thought of course that I would die within a few days. Mr. Carver immediately sent a very fine race horse and some money to purchase us back. The man (Bently by name) returned with the money and myself, having given the horse for me. The horse was valued at $300. Every effort was made to find my mother, but to no avail.

In the meantime my only two sisters died and were buried. My brother James and I grew up together, sharing each other's sorrows on the splendid farm owned by Mr. Carver. When just a mere tot in short dresses my very soul thirsted for an education. I literally lived in the woods. I wanted to know every strange stone, flower, insect, bird, or beast. No one could tell me.

My only book was an old Webster's Elementary Spelling Book. I would seek the answer here without satisfaction. I almost knew the book by heart. At the age of 19 years my brother left the old home for Fayetteville, Arkansas. Shortly after, at the age of 10 years, I left for Neosho, a little town just 8 miles from our farm, where I could go to school. Mr. and Mrs. Carver were perfectly willing for us to go where we could be educated the same as white children. I remained here about two years, got an opportunity to go to Fort Scott, Kansas with a family. They drove through the country.

Every year I went to school, supporting myself by cooking and doing all kinds of house work in private families. At the age of nineteen years

I went back to see my brother and Mr. and Mrs. Carver. I had not improved much in stature, as I rode on a half-fare ticket. The conductor
thought I was rather small to be traveling alone. I spent the summer here,
and returned to Minneapolis, Kansas where I finished my high school
work.

The sad news reached me here that James, my only brother, had died
with the small pox. Being conscious as never before that I was left
alone, I trusted God and pushed ahead. In working for others I had
learned the minutia of laundering. I opened a laundry for myself; got
all I could do.

After finishing high school here I made application to enter a certain
College in Iowa. I was admitted, went but when the President saw I was
colored he would not receive me. I had spent nearly all of my money, and
had to open a laundry here. I was liberally patroned by the students. I
remained here until spring and went to Winterset, Iowa, as first cook in
a large hotel.

One evening I went to a white church, and set in the rear of the house.
The next day a handsome man called for me at the hotel, and said his
wife wanted to see me. When I reached the splendid residence I was
astonished to recognize her as the prima dona in the choir. I was most
astonished when she told me that my fine voice had attracted her. I had
to sing quite a number of pieces for her, and agree to come to her house at
least once a week; and from that time till now Mr. and Mrs. Milholland
have been my warmest and most helpful friends.

I cooked at this hotel for some time; then opened a laundry for myself.
I ran this laundry for one year. This same Mr. and Mrs. Milholland encouraged me to go to college. It was her custom to have me come at the
day and rehearse to her the doings of the day. She would invariably laugh
after such a recital and say, "Whoever heard of any one person doing half
so many things."

She encouraged me to sing and paint, for which arts I had passionate
fondness. In one years time I had saved sufficient money to take me to
Simpson College, at Indianola, Iowa where I took art, music and college
work. I open a laundry here for my support. After all my matriculation
fees had been paid I had 10 c worth of corn meal, and the other 5 c I
spent for beef suet. I lived on these two things one whole week—it took
that long for the people to learn that I wanted clothes to wash. After that
week I had many friends and plenty of work.

I would never allow anyone to give me money, no difference how badly I needed it. I wanted literally to earn my living. I remained here for

three years; then entered the Iowa State College, at Ames, Iowa, where I pursued my Agricultural work, taking two degrees, Bachelor and Master.

After finishing my Bachelor's degree I was elected a member of the faculty, and given charge of the greenhouse, bacteriological laboratory, and the laboratory work in systematic botany.

Mr. Washington said he needed a man of my training. I accepted and came to Tuskegee nearly 27 years ago, and have been here ever since.

The 1922 autobiography was written at the request of Helen Milholland, the "prima dona in the choir" who sent her husband to bring young George home. Mrs. Milholland was at work on a biography of the up-and-coming scientist. Carver had no reservations about the worth of such a project. Already in 1920, with his most publicized years ahead of him, he penned this immodest letter to his would-be biographer:

12-1-20[4]

My dear Mrs. Milholland:-

I know you wonder what has become of me. Well your letter got misplaced in the rush of the opening of school. . . . Your letter turned up today.

I think if we wait awhile prices will drop to normal and we will be able to get it for almost 1/3 of the amount in fact I believe some publishers later will take it on a royalty basis.

If I can be of any service in looking the manuscript over for you and returning the same with suggestions I will be glad to do so, but frankly I want you to hold it permanently for should anything happen to me there would be a great demand for just such data as you have.

I have kept people here from writing one because I told them one was already written.

When I pass from earth to my reward there will be a great demand for such a book, and it will be a source of revenue for you, possibly in your declining years.

To give it more value I might give to it my endorsement.

I am also wondering if magazines would not pay for certain popular parts of it as an article that would be of interest to science or popular reading.

Much Love to all,

Kindly let me know what you think of what I have said.

Sincerely your friend

Geo. W. Carver

In 1922, a book entitled *Handicapped Winners* was written by Miss Sarah Estelle Haskin for the Methodist Episcopal Church South. It contained a laudatory chapter about Carver, who wrote to Mrs. Milholland, inquiring as to whether or not she had seen the book.

9-3-22[5]

My dear Mrs. Milholland:-

How glad I was to get your card, I have been thinking of you almost every day for months and since I did not get an answer to my last letter I concluded you had changed your address. . . .

I wonder if you have seen the little book entitled Handicapped Winners written by Miss Sara Estelle Haskin, Educational Secy. of the Board of Missions, Woman's work, M. E. South. It is written for their Schools, with the hope that it will bring about a better feeling for the negro.

It is published by their publishing house. I only have one copy or I would send you one. It has the "Peanut Wizard" in it.

Your name appears in it also. If you have a copy tell me please what you think of it.

How I wish you were nearer so I could come and see you once and a while.

My soul thirsts at times to see and have a chat with you.

God is blessing me more and more every day it seems.

I hardly think you would know me now I am quite gray, but pretty active yet.

Sincerely and greatfully yours,
Geo. W. Carver

Two weeks later, Carver again wrote to Mrs. Milholland about the Haskin volume. This time, he assured her that Haskin's treatment of his life was inferior to the work she was preparing.

9-17-22[6]

My dear Mrs. Milholland:-

How glad I was to get your good letter it seems so refreshing to hear from you. . . .

No the book does not even approach yours and I believe God in some way will provide a way for yours to come out, I am anxious for you to see this little book because it is to be put into their schools, for the information of the white boys and girls down here who seem to know so little about many types of colored people. . . .

Sincerely yours
Geo. W. Carver

By the late 1920s, Mrs. Milholland's manuscript had still not been published, although numerous other pieces about Carver had been. As always, Carver was unabashed in his comments about the worthiness of himself as a biographical subject and the continuing value of Mrs. Milholland's essay.

Jan. 16, 1928[7]

My dear Mrs. Milholland:-

I am attempting to answer your good letter with out having it before me. . . .

I am reluctantly sending you your most interesting manuscript. I have ten books with interesting sketches of my life on my desk now.

Nearly all the authors came and remained on the ground while writing them.

Yours however covers points they could not get.

I have never permitted anyone to see your manuscript. It is too unique I feared they would copy from it. Of course now it wants bringing up to date. From time to time I have tried to find the time to do it, but it seems impossible, I may be forced to retire before very long, as I am not very strong and am beginning to feel the weight of years.

I hope you will keep this manuscript so that I can revise it and make use of the unpublished facts it contains.

I do not feel so well tonight, so I will close and retire. . . .

Very sincerely yours
G. W. Carver

Less than two months later, Carver again wrote to Mrs. Milholland, encouraging her to maintain her interest in the project and again advising her that her material would be extremely valuable after his death.

Feb. 25, 1928[8]

My dear Mrs. Milholland:-

Thank you so much for your most interesting letter. Well with the manuscript, I thought you would give out just such information as you thought best. As I said before, you have facts there that no one else has, indeed you know of my early struggles as but few people do, and in other ways you know me as probably no one else does.

Another book just came to my desk a few days ago, with quite an autobiographical sketch.

By and by, before many years your manuscript will be most valuable as I will have passed on and then writers will be casting about to write more books, and make them complete. You have much of my early struggles and boyhood days.

Of course you was tired of it but save those fragments, they are valuable. . . .

I am so sincerely yours
G. W. Carver

Lucy Cherry Crisp, a long-time correspondent of Carver's, completed a manuscript about his life in the late 1930s. While searching for a publisher for the entire piece, she also tried to condense her story into an article for the *Saturday Evening Post*. The *Post* rejected her article, and Carver responded with two letters of advice for Ms. Crisp.

September 18, 1940[9]

My esteemed friend, Miss Crisp:

How happy I am to hear from you . . . .

I understand why the Saturday Evening Post did not accept your articles because the magazines have indeed been lavish in what they have had to say. I have been wondering whether it would be possible to get this into some foreign magazine, although the Orient is so torn up that I am not sure that it would have the value that it ought. I am thinking, however, that you will find an American publication that will be happy to get it, one that has not carried a write-up with reference to myself. It might be that some Southern publication will be glad to get it.

Most sincerely and gratefully yours,
G. W. Carver

Six weeks later, Carver penned another letter to Crisp, this time being more specific about how he thought she could increase the chances of having her manuscript published.

October 31, 1940[10]

My esteemed friend, Miss Crisp:

Thank you so much for your fine letter, the poems, and the article that you sent me. . . .

I have read the article entitled "The Man With Answers" and I believe that if you rewrite this article and cluster it around Henry Wallace some magazine and papers will handle it. I have received a number of requests from the political parties sponsoring the Vice-President, wanting me to write an article about our relationships as they felt that more knowledge should be given out with reference to Mr. Wallace's attitude toward the Negro and they thought I could do it better than anyone else. I have refrained from any political writing of this kind, and as I stated before, I believe your article worked over would meet a hearty reception.

This article, however, does not appear to me to have the same spirit of your other writings. I think one reason why the article has been turned down is due to the fact that in several ways it savors of the article published by Mr. Childers in the American Magazine, and while this article contained many good things and struck a popular chord with a certain type of people, the very careful thinker did not approve of a number of things that Mr. Childers dwelt upon such as how I looked, and how I dressed, and that sort of thing. Mr. Powell and several other white people with whom I talked from down town didn't like the article for that reason. I had a great many letters with reference to this article and not one of them mentioned that feature of it. I did get a few that regretted that he should put in all that.

I really believe that if you leave much of that out and put the spirit of your real genuine self into it that you will give us something quite different and find it readily accepted. . . .

Suppose that you try recasting this article and cluster it around my knowledge of Henry Wallace and our relationship. I am quite sure that you are thoroughly familiar with Mr. Wallace's attitude toward me.

I have just received an article about the same length as your's from President Gross of Simpson College. It is without doubt one of the most beautiful articles that has ever appeared regarding myself. He was here and spent several hours, and the article is high class from every angle and is rather different from any article that has appeared.

When you are yourself you can do just as well, and indeed you have a distinct and catchy characteristic in your own writing that no one else seems to catch. Think it over carefully and see whether I am right or wrong.

Most sincerely yours,

G. W. Carver

In the late 1930s, Mrs. Guy (Rackham) Holt began a biography of George Washington Carver. Working under contract with the publisher Doubleday, Doran, she visited Carver numerous times and corresponded with him extensively in an effort to gather resource material. The Carver who emerges from the pages of Holt's volume is a flawless hero of gigantic proportions. Holt's Carver suffers in silence, works tirelessly, thinks brilliantly, gives selflessly, and is imbued with such a sense of modesty that he spends much of his time in his last years parrying the compliments that his genius deserves.

Carver liked the way Holt portrayed him. It was his life as he wished it had been, not necessarily as it really was. The following letters reveal his intense interest in her book and his admiration for her portrait of his life.

July 23, 1940[11]

My dear Mrs. Holt:

I read the manuscript that you sent Mr. Curtis for me, and I want you to know that it is the most fascinating piece of writing that I have read. I started in and I confess I could not lay it down until I had finished it. You certainly have reflected your own personality into it in a most charming and delightful way, and the facts are really very clear.

Very sincerely yours,

George W. Carver

Subsequently, Mrs. Holt asked Carver to write yet another sketch of his early life, which, with the aid of his assistant Mr. Austin Curtis, he did. On 30 September 1941, he wrote again to Mrs. Holt.[12]

My dear Mrs. Holt:

This is simply to extend you greetings and to say that I hope you have the sketch of my early life that you wanted. Both Mr. Curtis and myself did the best we could with it.

I want to thank you also for the most fascinating article that appeared in the Christian Science Monitor some weeks ago. It is indeed illuminating and one that you do not want to put down until it is finished, then you wanted to reread it.

We are looking forward to your coming down before a very great while. . . .

I hope you will call on us for any additional material that you need, and it will be forthcoming if it can be gotten together. I have a book purported to be the only authentic information on the Carver family. Both Mr. Curtis and myself came upon a very egregious error, so much so that I am just a little afraid to do much quoting from it. For instance, it gives the purchase of my mother by Mr. Carver with her two boys. I have the bill of sale of my mother which shows that she was just a young girl when she was purchased, and according to that statement in the book, I would now be much over one hundred years old.

Most sincerely yours,
G. W. Carver

Mrs. Holt responded to this letter with a request for additional information about the details of Carver's early life and indicated that she would be traveling to Tuskegee soon to meet with him. Carver wrote back that he would try to help her fill the gaps in his life that she had been unable to fill, but that the trauma of growing up without parents had probably forced him to bury irretrievably much of his past.

October 13, 1941[13]

My esteemed friend, Mrs. Holt:

Thank you so much for your most interesting letter. . . .

I am happy to know that the book is making progress. I am sure that it will be unusually rich and fine. I shall be glad to assist you in filling in those gaps of which you speak. It will be very difficult indeed as there are so many things that naturally I erased from my mind. There are some things that an orphan child does not want to remember, but with such fragments as you have been able to collect, and Mr. Curtis, and myself, I am sure that we can piece together something that will not interfere with your high thinking too much.

With every good wish, and with the hope that I will see you before a great while, I am

Most sincerely yours,
G. W. Carver

Notwithstanding Carver's comment that there were things "an orphan child does not want to remember," he did sometimes speak of his childhood experiences. On March 18, 1939, Carver met with a group of about fifty people in the Curtis Hotel in Minneapolis, Minnesota. The gathering was orchestrated by Dr. Glenn Clark, a friend and biographer of Carver. Dr. Clark asked Carver to recount the story of how he found a much-sought-after knife when he was a boy on the Moses Carver farm. Carver told the story and, in doing so, revealed something more about his life as a boy with the white Carvers. Dr. Clark recorded Carver's comments.

> When I was a little boy, I wanted a knife. I wanted it in the worst way. I couldn't wait til I grew up. Little boys weren't supposed to carry knives. And I just couldn't wait twelve whole months to grow up! Mr. Carver used to go to bed when the chickens flew up into the trees to roost. My brother and I were supposed to stay in after dark. They told us stories of Raw Head and Bloody Bones[14] who would catch us if we went out. But my brother [Jim] and I used to go out to the persimmon tree. And then when we went into the house there was Mrs. Carver. And she always had a jar of willow switches. But when she started toward me, I used to open my mouth and yell. And she would walk away rather than have that awful yelling. And when she turned back, I would sob and stuff into my mouth a persimmon. And when I yelled I woke up Mr. Carver and he would call down, 'What are you doing to that pore chile?' And she would say, 'I ain't done nothin' yet'. But I could yell. I only got two lickings. That wasn't because I was so angelic, but I could yell, so you could hear me for miles.
>
> But to get back to the knife. One night I dreamed I was in the corn field, and in the middle of the row I found the rind of half a water-melon. I walked on 'til I came to the third hill of corn, and there was the other half of the watermelon, uneaten, with a knife in it. I woke up, and I couldn't go back to sleep. In the morning after breakfast, as soon as it was light enough so that Raw Head and Bloody Bones had gone, I started out. I walked straight to the corn as though I was being led. There was the rind. And then I walked to the third hill, and there was half a watermel-on, with the knife sticking in it. It was a knife with a black handle—what we called the Lady's knife. And it had two blades. I took that knife with me when I went to Tuskegee forty-three years ago. . . . [15]

Carver encouraged Holt's progress at every opportunity. As the book neared its completion, he became excited about its reception by what he anticipated would be a wide readership.

August 4, 1942[16]

My esteemed friend, Mrs. Holt:
    This is to thank you for your letter of recent date which I am very glad indeed to get. . . .

Your letter is quite refreshing. I am glad that you are able to get in as much of the manuscript as the printers can use at the present time, as the calls that are coming in for it are simply remarkable, and the sooner it gets out of my mind the better.

You will notice that the publications with reference to the little log cabin etc., that have come out recently are unusually good, and just as soon as I can get the publication matter from Doubleday Doran Company, I believe the book will receive a storm of applause. . . .

I want you to know also that I have never in all my experience had a Bulletin to receive the real ovation like the Bulletin on "Weeds".

With every good wish, I am

> Most sincerely and gratefully yours,
> George W. Carver

In the fall of 1942, Carver was informed by a Doubleday, Doran executive that Holt's biography would be out early in 1943. He responded to this information with a letter about the widespread demand he believed there would be for the story of his life.

> October 13, 1942[17]

My esteemed friends, Mr. and Mrs. Woodburn:

This is to thank you for your most interesting letter. . . .

I note that most of the book is now in, and that you hope to be able to announce its issue early in 1943, that is, you hope to go to press before Christmas.

I am very certain that the book will be one of the outstanding biographies that has come out recently. I believe that every soldier will want a copy.

My strength comes and goes at times. I am able to get to my office occasionally. At other times I am absolutely unable to get there at all.

With every good wish, I am

> Sincerely yours,
> G. W. Carver

Although Carver was pleased to learn that Holt's book would finally be available to the public, he was not happy that its release was still months away. Soon after his letter to the Woodburns was written, he received correspondence from Mrs. Holt in which she asked for additional material. Carver was distressed over Mrs. Holt's request. By this time, his waning health was of real concern to him. He wanted the book to come out while he was still alive, and he was impatient with delays. On 14 October 1942,[18] he wrote what turned out to be his last letter to Rackham Holt. The "something sordid" to which he refers is, of course, the possibility of his own death.

My dear Mrs. Holt:

Thank you very much for your interesting letter and the material which you returned. . . .

I wish so much that the book could be finished. I think that it should be closed up. I had a letter from Mr. Woodburn stating that it would be sometime in December before the book could be announced. I was hoping so much that this book could be finished before it had to close with something sordid. . . .

With reference to the paintings, I thought that I made it clear sometime ago that I myself had absolutely nothing to do with the paintings or anything pertaining to the gift. They are the property of the Foundation governed by a Board of Trustees of which President Patterson is Chairman of the Board. . . . They have taken the firm stand that nothing must be removed from the galleries, and nothing can be removed without the action of the Trustees. I, who established the Foundation, would naturally not have anything to do with it except by courtesy, and that is the reason I am on the Board of Trustees, and will be as long as I live. I go to the meetings when I can; otherwise I do not, so that there is absolutely no chance of getting the pictures.

Now as to the last paragraph, Mr. Ford will not be in Detroit at the time you designate, I am sure, as he comes South early in the spring to his Richmond Hill estate. He hasn't designated to me as yet just when this trip will be made, but he stops to see me either going or coming as we have many matters to take up in connection with our work, so he stops over for about two hours. All the time, of course, except a few moments is closeted with myself.

Now may I suggest that you get in touch with Mr. Frank Campsall and make your plans to visit Dearborn and the Edison Institute. Mr. Campsall will naturally direct you as far as they think wise, and naturally you will be directed to the Engineering Laboratories with Professor R. A. Smith in charge. Mr. Ford is naturally a very difficult person to see, and then unusually reticent. . . .

Mr. Curtis is away, and has been for nearly a week, so I am here struggling along as best I can, so may I urge again to please finish the book as I would like to have it come out while I am still able to see it, and not have it, as I stated before, to end sordidly which must be done pretty soon.

With every good wish, and with regrets to write a blue letter like this, yet it seems quite necessary for your sake as you have what no other person in the world has, as I told Mr. Curtis last week that you have material that he himself has not, and it would take a careful writer many months such as you have spent in getting this material. Please let us close it out as rapidly as possible.

Sincerely yours,
G. W. Carver

In the late 1930s, Glenn Clark, a literature professor at Macalester College in Minnesota, wrote a pamphlet titled *The Man Who Talks with Flowers*. Clark was a layman who was very interested in religion and who wrote a number of religious tracts. *The Man Who Talks with Flowers* was based upon Clark's correspondence and conversations with Carver over the course of more than a decade. In the early 1940s, another mystic, Gloria Dare, wrote to Carver in an effort to get him to collaborate with her on his life's story. He responded by referring her to Glenn Clark's book and telling her that he was incapable of writing about himself—he preferred others to do it.

August 7, 1941[19]

My beloved friend, Miss Dare:

I have read your letter as usual with much interest and satisfaction. I presume you have a copy of Dr. Glenn Clarks little booklet on "The Man Who Talks With the Flowers", it is just a simple account of his two visits here and he tells it in his own imitable way. The book is being sold by the thousands, it is perfectly remarkable the people who are sending for it. I understand that book stores are ordering it in large quantities so that they may have a supply on hand at all times. I have a letter to that effect from the publishers and always everyday letters come to me with reference to it.

It seems to me like anything that I would write would fall so short of the beautiful way that you put things that it would detract from it rather than add anything and then to me it is so refreshing to get just what people of real worth say. Not to please me however, but the real estimate of worth. Please think this over and if you agree, we will let it go just that way.

I believe that your writings would have so much more weight and if you know how impossible it is for me to write anything about myself because the more I attempt to write about myself, the worse it gets and after while I get so disgusted with it that I just throw it away.

I am most gratefully and

Sincerely yours,
George W. Carver

The above letter indicates how impressed Carver was with Clark's work. He suggested it as a model to his friend Lucy Cherry Crisp, who was still struggling to get her manuscript about his life published.

August 14, 1941[20]

My beloved friend, Miss Crisp:

Thank you so much for your very interesting letter which has just reached me.

I am glad to know that you are going to proceed along the lines with reference to your book as you have applied it as that is the only way to get the best results. I did not mean to convey the idea at all that you should change the book, but I did feel and do yet, that you could write a small booklet that would be quite as catchy and popular as Dr. Clark's as you two have such a unique way in expressing yourself. That is, I have not had anyone express themselves just as you do, and naturally, a small booklet would not be in any since a replica of the larger one. I do appreciate, however, the comments made by your friend. He brings up several angles that I have not thought of myself as I am not an authority on book writing or salesmanship either. . . .

With every good wish and highest personal regards, I am so

Sincerely yours,
George W. Carver

In 1942, less than a year before his death, Carver was still encouraging Miss Crisp to finish her book, telling her that hers would be different from the one being prepared by Rackham Holt.

March 11, 1942[21]

My esteemed friend, Miss Crisp:

This is just to extend you greetings, and thank you for your letter. . . .

I am glad you take the attitude towards your book that you do, as it will be one entirely different from the one Mrs. Holt is writing. She is here yet and will probably be here for some weeks longer. The Doubleday

Doran people seem to not care at all with reference to how much money they are spending on this book. It must be costing them thousands of dollars because to send her around over the country and keep her here for months at a time as she has been more than two years writing this book. Your's will be entirely different and will sell at a much smaller price, which to my mind will meet the pocketbooks of the layman. I fear as elaborately as the Doubleday Doran people are illustrating their's the price will be beyond the poorer people. Nevertheless, if that is what they want, they are the publishers and putting their money into it so they know much more in five minutes than I would in five years because I am not a publisher, and they are. . . .

Very sincerely yours,
G. W. Carver

Despite his glowing praise of adulatory articles and books written about him, there was another side to Carver. At times he could be very self-effacing, particularly when he attributed his success and creativity not to himself, but to the power of God working through him. At times, too, he could be downright humble, telling his correspondents that the nice things they said about him were really overstated. How much of this humility was genuine and how much was false, designed to foster an image of himself as the humble genius, is impossible to say. One gets the impression, in reading many of his letters, that he, like most people, really did like it when people praised him, but that on occasion felt that etiquette demanded at least a partial disavowal of his brilliance. In the following brief letter, for example, he acknowledged that an article about himself was a "fascinating story" but went on to say that he did not deserve all of the credit he was being given.

3-21-32[22]

My beloved friend, Mr. Zeller:-

Before me is a copy of "The Winterset Madisonian", containing your wonderful article. I say wonderful advisedly.

It is an article that never looses its charm, it is rich in thought, correct in what it portrays, and is truly a fascinating story, when one starts to read, they are loth to put it down until finished. I do not deserve all the credit you have given me. . . .

May God continue to bless and keep you always.

Admiringly yours,
G. W. Carver

At times Carver, like many a celebrity, seemed to complain of the attention he received and the way in which it altered his lifestyle, but his protests were often more apparent than real, as in this letter to M. L. Ross, a Topeka physician and long-time friend.

January 27, 1942[23]

My beloved friend, Dr. Ross:

This is just to extend you greetings, and to say that I had the pleasure of seeing Mrs. A. B. Griggs for just a few minutes. She came unannounced, and there are some days when I cannot see anybody at all according to the doctor's orders.

For about three weeks I have had an attack of "flu" that robbed me of my strength almost completely and took my appetite, and naturally I got so weak I could hardly stand up. I am just now beginning to get my physical strength back again, but you know how our people do very largely. They wait until it is too late, and then make a grand rush to see you, take your picture, news writers want your story, artists want you to pose, and all manner of things I can't possibly do.

In addition to that I have a person who is writing a book and she is here and has been several times, and I am obliged to see her daily if I have the strength to do it. This book will be issued by the Doubleday Doran Company and will doubtless be a monumental volume as she is considered one of the best writers in the entire country. . . .

With much love, I am

So sincerely yours,
G. W. Carver

Indeed, if Carver could be somewhat vain when talking about the praise heaped upon him, he could also be very humble, especially when acknowledging that the real source of his power and success came from God, as in this letter to one of his "boys," Jimmie T. Hardwick (see Chapter 9).

5-21-24[24]

My very dear friend, Mr. Hardwick:

Your fine letter reached me this evening and I cannot forego the pleasure of writing to you right away.

I am glad you enjoyed your visit here. . . .

My friend, God has indeed been good to me and is yet opening up wonders and allowing me to peep in as it were. I do love the things God has created, both animate and inanimate. As he speaks aloud through both, God willing, at Blue Ridge we will let Him talk to some of us.

You do me too much credit. I am not so good. I am just trying, through Christ, to be a better man each day. Your spirit helps me so much. It is what my very soul has thirsted for all these years, a spirit that God likewise was developing to perform a great service to humanity, such as he is developing in you. . . .

My "horoscope" tells me that God is yet going to do some thing for you that will astonish you. Just to think you have been here. We are to see each other in June, and this is not all. He has something else in mind for us. My prayers and blessings are always for you.

> Sincerely yours,
> Geo. W. Carver

One of Carver's favorite tactics of self-deprecation was to tell friends who had praised him that they had overstated their case by about 75 percent. Yet it is obvious that Carver enjoyed the attention he drew. He was fascinated by his own life story; he reveled in the fact that people traveled to Tuskegee just to get a glimpse of him; and he never tired of encouraging would-be biographers to turn out still another account of his well-publicized accomplishments. P. H. Polk, the famous Tuskegee photographer, recalled in 1974 that Carver had told him, "Mr. Polk, let me tell you something. Every opportunity you get, you make my picture because when I'm dead folks going to want them."[25] Likewise, Carver "boy" Dana Johnson remembered in a late-life interview: "He [Carver] told us a number of times 'save my letters and these papers and things that I give you because some day you will be glad that you did.' He had this feeling I think right from the start that he was a man of consequence."[26] Carver may have felt that much of what he accomplished was the result of a Supreme Being's work through him; but he constantly looked to the world of man for confirmation of that fact.

THREE

The Pre-Tuskegee Years

Old Friends Remembered

You, of course will never know how much you done for a poor colored boy who was drifting here and there as a ship without a rudder. You helped to start me aright and what the Lord has in his kindness and wisdom permitted me to accomplish is due in a very great measure to your real jenuine Christian spirits, how I wish the world was full of such people, what a different world it would be.

Geo. W. Carver 2 September 1901

GEORGE WASHINGTON CARVER never forgot his friends, particularly those who assisted him in his early days as a wanderer and a student. Throughout his long life he maintained correspondence with dozens of people who had affected his life in the years before he arrived at Tuskegee.

There is little extant correspondence between Carver and the people he encountered before leaving Missouri as a young boy. Neither is there any evidence that he returned to Newton County more than a few times. Yet the area, and the memories it evoked, always seemed close to him. In the late 1920s, Carver began corresponding with Eva Goodwin from Neosho, Missouri, the daughter of Thomas Williams, a greatnephew of Moses Carver, George's master. Williams and Carver had been boyhood playmates, despite the fact that the former was several years older. Carver's fondness for the place of his youth, and the "homefolks," as he called them, is obvious in this letter.

Dec. 28 - 28[1]

My esteemed friend and home-folks, Mrs. Goodwin:—

What a wonderful holiday greeting this is I consider the finest of all.

Your letter causes tears to come to my eyes as I recall childhood's happy days.

As a boy, Thomas Carver, (I called him, your Father) was my ideal young man, I loved him as I did my own brother, and indeed he was a very dear handsome young fellow.

How I would love to see him, I am sure neither would know the other, as both have changed so much in all of these years.

As I remember it you was just a little girl when I left. It is a little hazy in my mind as to how you looked, if you were even born.

I passed through Neosho last Oct. a year ago enroute to Tulsa Okla. Where I had the peanut exhibit displayed at the State Fair.

The train stopped for quite a little while at Neosho, but it was dark and I could recognize nothing.

The New Junior High School which is being erected for colored children is named for me and the Board of Education is very anxious that I come out and dedicate it.

It will be sometime this coming Feb. I am not sure if I can make the trip as I am not very strong now.

I certainly feel just as you do about kinship.

Your letter over comes me with Joy, I cannot write any more now, this letter must reach you before the New Year comes in.

Please let me hear from you further. The address is correct.

So sincerely yours,

With love to all my dear home folks.

Geo. W. Carver

Eva Goodwin responded to Carver's letter with a photograph of her father and a newspaper obituary of Moses Carver, who had died in 1910. The professor answered warmly.

Jan. 25 -29.[2]

My dear Mrs. Goodwin:—

It is impossible for you to know how your letter made me feel, if you had only been a hundred mile or so away I would have started immediately to see you.

I have sat and looked long and hard at your Father's picture. While he has changed quite as much as I have, I can still discern that handsome kindly face, which made him to me ideal.

Of course you are happy to do the things for him, how I love this picture, one of my dearest boyhood playmates.

If I go to Oklahoma, the way I went before I will pass through Neosho, and it may be that I can arrange to stop over a day, or between trains at least. And see you dear Father and the rest. I am too beginning to feel the weight of years and cannot do much traveling now.

How I would love to get with your Father and talk over old times at home, indeed you really are my home folks.

Thank you for the clipping referring to the death of dear Mr. Carver, (Uncle Mose) I treasure it very much.

How delightful to have you speak of your Father in that way. I beleive every word of it, his face shows it.

Yes I can remember you as a little girl, used to hold you on my lap. I certainly would appreciate any pictures of the old homeplace. I am sure it has changed very much.

I thank the good people for their words of approval, why should I not be able to do pretty well, I certainly had good home training by my "home folks".

My heart indeed goes out to my dear "Home Folks", Love and the best of wishes to your good neighbors.

I am so sincerely yours,
G. W. Carver

In August 1886, Carver homesteaded one-hundred-sixty acres in Ness County, Kansas. He worked the land for approximately two years before drought and a lack of capital forced him to abandon the venture and move on to Iowa. During those years he was befriended by many people, virtually all of them white. In the early 1930s, an article about Carver reached George A. Borthwick, who had once loaned him money. Borthwick wrote to Carver, asking if he was the same George Washington Carver who had lived so many years before in Kansas. Carver responded as follows.

Oct. 16-32.[3]

My dear Mr. Borthwick:—

How your letter astonishes me. Yes I am the same Geo. W. Carver.

I do indeed remember you so pleasantly, when you came to the Steely home.

My, how I would love to see you and those other dear boys as well as some of my old haunts.

Glad you liked the story in the American, cut it down 75% and you will get at just about what is right.

The enclosures may interest you.

So sincerely yours,
G. W. Carver

Following that letter, Carver began corresponding with Borthwick and other residents of Ness County, including O. L. Lennen, a county official. Lennen and Borthwick wrote articles about Carver in local newspapers, and in 1942, less than four months before he died, Carver penned this note to Lennen.

September 24, 1942[4]

My esteemed friend Mr. Lennen:

This is to acknowledge receipt of your fine letter which came yesterday.

I want you to know also that I received the extra papers, for which I am very grateful. Of course every one who sees the paper wants it, but I cannot let them have mine, as I have certain strategic points into which these papers will fit. I wish to thank you for the fine article, not only for the historical data, but for the personal interest you take.

I am more and more enthused over the fact that you and Mr. Borthwick must come down, and see some of the things about which you have been writing, and some of the things the public will not believe because they cannot conceive of such things, but they are here. It also emphasizes the type of mind that you and Mr. Borthwick had, that you were able to look into the future of a struggling young Negro boy and discern that there was something in him worthy of a chance in life like other folks. If you could only take a day or two off and study the Museum alone, I am sure it would open your eyes even wider.

The narrative you gave the young man is essentially correct.

I hope you will talk it over with Mr. Borthwick and see if there is not some way by which you can both get down here. . . .

Most sincerely yours,
George W. Carver

Another Ness County native with whom Carver corresponded was the editor of the county newspaper, Mr. Knox Barnd. In 1935 Barnd wrote an account of the Old Settlers Reunion in the *Ness County News*. He also featured an article about one of Ness County's most famous ex-residents, George Washington Carver. Carver responded with this nostalgic note.

June 10, 1935[5]

My dear Editor Barnd:

I presume that I am endebted to you for a copy of June the 8th issue of the News giving such an illuminating account of the Old Settlers Reunion. As I looked over the program, I grew very much enthused. I wish so much that I could have been there to have shared in it. I saw several names that I remember, and am certain had I been there I would have seen quite a number of my old friends.

Another interesting and pleasant thing to note is the improvement in the paper. I am amazed at the size of it, appearance, and the splendid news items written in a way that if one starts to read an article he is loathe to put it down until finishing it.

There is only one article that I see that could be improved, and that is the one entitled "George Washington Carver", which is overestimated seventy-five per cent.

I want to say, however, to the good people of Ness City that I owe much to them for what little I have been able to accomplish, as I do not recall a single instance in which I was not given an opportunity to develop the best that was within me.

In looking over the bird's eye view of Ness City, it seems that I can remember some of the houses. Should I ever be priviledged to come that way, expect me to stop.

Again thanking you for the paper, I am

Yours very sincerely,

G. W. Carver

Two of Carver's most cherished acquaintances from the pre-Tuskegee days were John and Helen Milholland of Winterset, Iowa. Carver ended up in Winterset in 1890, more by chance than by design, after traveling around the Midwest for roughly a decade, spending most of his time in Kansas. The Milhollands met Carver at church and invited him home to dinner. Subsequently, Carver became a frequent guest in the Milholland house. He had aspired to go to college, but had recently had a most unpleasant experience with a small denominational school in Highland, Kansas. He had been accepted for admission by mail, only to be denied entrance because of his color when he arrived to begin classes. The experience left him dispirited and afraid to try elsewhere. Mrs. Milholland, in particular, recognized the young man's latent genius and encouraged him to continue formal study in the area that seemed to give him the greatest joy: art. Carver agreed to try enrolling at nearby Simpson College; he was accepted, began attending, and was always grateful to the Milhollands for their encouragement.

Carver's student days were no doubt trying—he had few resources to fall back upon, little money, and no family. He was unquestionably gifted and just as unquestionably hungry for acceptance by his fellow students. As was so often the case with the complicated Carver, his letters revealed his simultaneous aversion to and fondness for self-praise. Understandably, he wanted the people who had helped him to know that he was doing well. The following letter, the earliest available correspondence by Carver to the Milhollands, shows not only that side of the young man, but also the diversity of his interests.

[Undated][6]

Well Mr. and Mrs. Milholland here it is nearly 11 oclock sunday evening of the next week and your letter not finished, I hope to be able to finish it this time I have been trying to write ever since I received you letter as I was so very glad to hear from you have been quite sic since you wrote and had to be excused from school one afternoon but I am better now I get a scolding nearly every day about working so hard by my teacher we will have vacation in a week for one week I want to go up to Des. Moins during that time. I have been waiting to paint or rather get an idea of the transparent work in flowers before sending you the copy I am painting flowers now I have nearly finished one very large marine and am painting on an origion design of the cactus and yucca like my teachers only very much larger the canvass is 22 x 48 inches The people are very very kind to me here and the students are wonderfully good they took into their heads that I was working too hard and had not home comforts eneough and they clubbed together and bought me three real nice chairs and a very nice table they left them for me while I was at school. I have the name unjustly of haveing one of the broadest minds in school. My teacher told me the other day that she is sorry she did not find me out sooner, so she could have planned differently for me Well this subject is getting very monotonous to me and was before I begun it, but I thought you would like and be interested in knowing what they thought of me; please dont let any one see this letter but the home folks. . . . Nearly 12 please excuse me and write sooner than you did before My best respects to all I remain your humble servant of God I am learning to trust and realise the blessed results from trusting in Him every day I am glad to hear of your advancement spiritualy and financially, I regard them also as especial blessings from God

Sincerely yours

Geo. W. Carver.

As in the following letter to the Milhollands, Carver often wrote of his feeling that God had chosen him for a special mission in life, although, for the moment at least, he was uncertain just what that mission was to be.

April 8, 1890[7]

Mr. & Mrs. Milholland, dear friends:—

Since you wrote, me I have been to Des Moines and back again It was very muddy, but I had a nice time, I was to see your sister Mrs. Maley and took dinner with her. . . . I am taking better care of myself than I have, I realize that God has a great work for me to do and consequently I must be very careful of my health. You will doubtless be surprised to learn that I am taking both vocal and instrumental music (piano) this term. I don't have to pay any direct money for any music, but pay it in paintings they heard me sing several times and then they gave me no rest until I took it or rather consented to. They are very kind and take especial pains with me. I can now sing up to high D and 8 octaves below I have only had one lesson. He told some of the students that he thought he could develop some wonderful things out of my voice or something to that effect. . . .

I am glad the out look for the upbuilding of the kingdom of Christ is so good. We are having a great revival here. 40 seeking last night and 25 rose for prayers at the close of the service. I did not go tonight. My best regards to all and to your nephew also as I suppose he is there now Shall be glad to hear from you soon.

Geo. W. Carver

In 1891, Carver enrolled at Iowa Agricultural College (later Iowa State College) to pursue an education and a career in the more pragmatic field of agriculture. His first letter to the Milhollands after arriving in Ames reveals that he did not like it as well there as he had at Simpson. He had come to believe, however, that God's plan for him included studying subjects that would be of greater utility to African Americans, so he resolved to persevere.

Aug. 6, 1891[8]

Mr. & Mrs. Milholland, dear friends:—

I will try to pen you a few lines in answer to your very kind epistle of a recent date. . . . I as yet do not like it as well here as I do at S. because the helpfull means for a Christian growth is not so good; but the Lord helping me I will do the best I can. . . . I am glad to hear that the work for Christ is progressing. Oh how I wish the people would awake up from their lethargy and come out soul and body for Christ. I am so anxious to get out and be doing something. I can hardly wait for the time to come. The more my ideas develop, the more beautifull and grand seems the plan I have laid out to pursue, or rather the the one God has destined for me. It is really all I see in a successful life. And let us hope that in the mysterious ways of the Lord, he will bring about these things we all so much hope for. I wish it was so that we could assist each other in the work as there is such a sameness in it, and I seen by one of the late southern papers that one of their strongest men advocates a broader system of education, and lays down a plan very much like the one I have but not as broad. And the more I study and pray over it, the more I am convinced it is the right coarse to pursue. . . . Let us pray that the Lord will completely guide us in all things, and that we may gladly be led by him. . . . My hope is still keeping without becoming stale either.

George

Although Carver had initially expressed dissatisfaction with life at Iowa State and measured it unfavorably against his experiences at Simpson College, he soon adjusted to the larger school. By the time he wrote the following letter in 1894, he obviously felt gratified at being an integral part of the Iowa State family.

Oct. 15th-94[9]

My dear friends:—

I was indeed more than pleased to hear from you. Had you not written you would have received a letter from me soon as I intended writing. . . .

I am glad to know that you are all well and that the Lord is blessing you so unsparingly. Beg pardon for finishing with a pencil but my pen has run dry and I have no ink with which to fill it. The Lord is wonderfully blessing me and has for these many years. I cannot begin to tell you all I presume you know I had some paintings at the Cedar Rapids Art exhibit, was there myself and had some work selected and sent to The World's Fair, was also sent to Lake Geneva twice to the Y.M.C.A. Summer School as a representative from our college.

And the many good things the Lord has entrusted to my care are too numerous to mention here. The last but not least I have been elected Assistant Station botanist, I intend to take a post graduate course here, which will take two years. One year of residence work and one of non-residence work. I hope to do my non-residence work next year and in the meantime take a course at the Chicago, academy of arts and Moody institute. I am saving all the pennies I can for the purpose and am praying a great deal. I believe more and more in prayer all the time.

I thank you very much for your kind invitation to visit you nothing would give me greater pleasure and I will do so if I can but I will have to cancel some alter cantemplated visits. There are so many that are so determined that I shall make them a visit this winter. But I think I will get to see you as I have much to tell you and ask your advice about.

I have been to A reception nearly every night last week and the cards are out for part of this week. One given by the Pres. of the College and several down town that are not connected with the institution I dont pretend to go to all of them. . . .

Enclosed please find catalogue and classification of our Miss. library which has been secured and paid for this year, our band 2 yrs. ago was organized as a reading circle with 3 members including myself. We now have a num. of about 29. with 2 volunteers God bless you

Geo. W. Carver

Carver left Iowa to take a job at the Tuskegee Institute in the fall of 1896. Soon after he arrived at Tuskegee, he penned a "thank you" note to the friends he left behind at Iowa Agricultural College. The note appeared in the IAC newsletter, October 27, 1896, and bore a sound of disappointment in the circumstances he found at Tuskegee:

Oct. 22, 1896[10]

My Dear I.A.C. Friends:

This evening as I sit at my writing desk, many miles away from you, in the sunny south-land, I wish I could make you feel how thankful I am for the beautiful and useful presents you so kindly gave me: You who have no such problems to face as I have here can scarcely appreciate their usefulness to me. There is nothing I needed more. May that kind Providence who directed you to give me these things reward you so abundantly is the wish of your friend, student and classmate,

Geo. W. Carver

P.S. I am well pleased with my work and will write you a long letter as I can spare the time.

Within a few years of arriving at Tuskegee Institute, Carver had become a fixture at the Alabama school. Still, he continued his correspondence with the Milhollands, as well as with others at both Simpson and Iowa State, and he remembered the kindnesses that had been shown him. He wrote to the Milhollands in 1901:

9-2-1901[11]

My dear freinds, Mr. and Mrs. Milholland,

Your esteemed favor of July 23rd reached me a few days ago and I was indeed glad to hear from you. . . .

I think of you often and shall never forget what you were to my life, how much real help and inspiration you gave me. You, of course will never know how much you done for a poor colored boy who was drifting here and there as a ship without a rudder. You helped to start me aright and what the Lord has in his kindness and wisdom permitted me to accomplish is due in a very great measure to your real jenuine Christian spirits, how I wish the world was full of such people, what a different world it would be.

Should I ever come so near you again rest assured I am coming to see you, and I plan to come to Iowa during some of my vacations. . . .

Must close as I have about 20 letters to answer aside from my other duties.

May God continue to bless and keep you.

Love to all.

Yours very greatfully
Geo. W. Carver

In 1914 Carver was in an automobile accident that left a companion seriously injured. He wrote to the Milhollands about it and also offered some comments on the Great War then raging in Europe.

Dec. 23d-'14[12]

My dear freinds:—

How I enjoyed your delightful greeting. You all are so often in my mind, as the years come and go.

I am glad to know that all are well, and am especially thankful that the good Lord has spared me to write to you.

This summer I came near loosing my life, and I am yet unable to see how I could pass through such an ordeal and yet live. A large auto truck turned turtle with several of us in it, one man was badly mashed up so much so that he is yet after 7 months unable to walk, I was pinned down under the truck, badly bruised and cut up but no bones broken.

Every time I pick up a paper I think of what Mr. Sherman said war was, words fail to describe the horror and suffering.

It is making it very hard here for us, not much money is coming in and how we will get along God only knows.

I have just learned that 25 girls are nearly barefooted and have not sufficient closes to keep them decent or warm.

Many of the boys are almost as bad, so we are going to do what we can for them.

I do hope the entire holiday season will bring to you many Joys and permit you to share it with the less fortionate.

Sincerely your friend,
Geo. W. Carver

No one influenced Carver's scientific training more than Prof. L. H. Pammel of Iowa State College. Carver acknowledged this reality in 1918, when he wrote this to Pammel: "I certainly consider it an honor to have been a pupil of yours, and as I have said to you a few times and to others many times, you influenced my life possibly more than anyone else."[13]

Pammel was a mycologist, an expert on fungi and plant diseases. He was among the most prominent botanists in the country, evidence of Carver's exemplary scientific training. Pammel and Carver were kindred spirits. Professor Pammel, as historian Mark Hersey points out, published the first book in English with the word "ecology" in its title: *Flower Ecology*, which appeared in 1893.[14] Carver took many courses with Pammel as an undergraduate, and then, as a graduate student, conducted research under the direction of the famous scientist. Years later, Pammel would remember Carver as "the best collector I ever had in the department or have ever known."[15] Carver was equally gracious in his description of Pammel. They published several scientific articles together and were still collaborating after Carver began his Tuskegee career.

6-5-1899[16]

My dear Prof Pammel

Your esteemed favor of May 29th at hand and contents duely noted. I hope you will pardon me for giving you so much trouble concerning my paper for the Acad. of Science.

I fear I could not get up a paper worthy of presentation on puff balls this year as I have only a very limited number in my collection and have given them but little thought since I came, again here they do not appear until quite late in the fall, making it quite late to collect and identify.

I am making a specialty of the circos form have a set nearly ready to send you, nearly all or quite all are named by good authorities and will add more to the list. I should like [illegible] & Macon Co. If you will help me with it, I have never been anywhere or anyplace whence Cir. were so abundant. I appreciate your effort to help me, but this is not new, or unusual to me, as during the 6 years I was privileged to be under your tutor ship and at times when I know it must have cost you much to do it. It does me a great deal of good as I read from time to time clippings of your success. . . .

Yours most truly
Geo. W. Carver

In 1921, Pammel visited Carver at Tuskegee. Carver was elated, and after Pammel had returned to Iowa, he wrote to his mentor about how much he appreciated the visit. Carver also thanked Pammel for numerous gifts the latter had given him, including a microscope.

6-11-'21[17]

My dear Dr. Pammel:-

I am just recovering from the extreme pleasure of having you visit me, the fulfillment of a long cherished hope.

I had talked so much about you that everyone was eager to see you, and said without hesitancy that I ought to know something coming up under such a scholar as yourself.

My only regret was that you could not stay longer and that school was not in session.

Everyone wanted to hear you speak. I really wanted you to see the suit of clothes, hat, gloves, underwear, you helped fool me down town and bought for me, preparatory to going to Cedar Rapids to the Art exhibit with some of my pictures.

All of the things mentioned are nearly as good as new, the gloves show the most wear.

Of course my mycroscope is just as good as new.

I was so glad to see that God had dealt so kindly with you, by giving you increased bodily vigor, great mental attainments etc.

When you were going out of your way to help a poor insignificant black boy, you were giving many "cups of cold water" in His name.

The memory of yourself, Mrs. Pammel and the children are more dear to me than my words can express.

They have served as lamps unto my feet and lights along my pathway. . . .

Sincerly and greatfully yours,
G. W. Carver

In the early 1920s, Professor Pammel was preparing a brief article on Carver's life. He wrote to his former student and asked for the latter's recollections of his student days. Carver responded with the following letter, a summary of his memory of people and events from twenty-five years earlier.

May 5, 1922.[18]

My dear Dr. Pammel:

In response to your queries of recent date, I beg to reply as follows:

1st. Born at Diamond Grove, Mo., just as freedom was declared, in a little one roomed log shanty on my master, Moses Carver's farm.

2nd. My education was picked up here and there. Mr. and Mrs. Carver taught me to read, spell and write just a little. I went to Neosho, Mo., public school for about nine months, then to Fort Scott town school, for about the same length of time. From there, I went to Olathe, Kans., where I attended the town public school for about two years.

Leaving here, I went to Minneapolis, Kansas, where I nearly finished my high school work. From here, I went to Indianola, Iowa, to Simpson College, where I took the College work and specialized in art and some music.

From here I went to Ames, Iowa to take a course in Agriculture, persuaded to do so by my art teacher, Miss Etta M. Budd, to whom I am greatly indebted for whatever measure of success that has come to me.

Miss Budd helped me in whatever way she could; often going far out of her way to encourage and see that I had such things as I needed.

During my six years in College, her interest in me never waned.

3rd. I do not now recall the exact date.

4th. I did odd jobs of all kinds for a number of the professors, such as cutting wood; making gardens; working in the fields; helping clean house; taking care of the green house and the chemical, botanical and bacteriological laboratories.

5th. Came to Tuskegee Institute, and took charge of the Agricultural Department here; kept it about fifteen years, then was given charge of the Agricultural Research work. I have kept this work in connection with the Experiment Station ever since.

6th. I have no words to adequately express my impressions of dear old I.S.C. All I am and all I hope to be, I owe in a very large measure to this blessed institution.

7th. "Beardshear", was one of the biggest and best hearted men I have ever known and it was so pleasant and uplifting to come in contact with him.

"Wilson", the name of Hon. James Wilson is sacred to me. He was one of the finest teachers that it has ever been my privilege to listen to. He taught a Sunday School class in which every student would have enrolled, if they had been allowed.

The class grew so large that he conceived a very unique plan to divide it, so he graduated some twenty or twenty-five of us who had been with him the longest and to some he gave classes. I happened to be one of the ones graduated. We all left him sad and reluctantly. We gave him to understand, in no uncertain terms, that we did not like it at all and out of our love for him, we went, but in less than two months we were all back again.

Our displeasure grieved him very much and he said to me, many times that he would never try to divide his class again, no matter how large it got.

Being a colored boy, and the crowded condition of the school, made it rather embarrassing for some, and it made the question of a room rather puzzling. Prof. Wilson said, as soon as he heard it "Send him to me, I have a room," and he gave me his office and was very happy in doing so.

"Budd" was the father of Miss Etta M. Budd, heretofore mentioned, and my professor of Horticulture and a man much on the order of Prof. Wilson; kind, considerate, loving and loveable; a great teacher, and he made of his students his personal friends.

Everybody loved Prof. Budd.

Stanton, "Stanty", as he was affectionately called, did all within his power to enthuse, and inspire me in the science of plane geometry; but, having such poor material to work with, his efforts were not crowned with very brilliant success, which of course was not his fault.

Prof. Stanton was the finest teacher of Mathematics I have ever seen. He was universally loved by every one who had the privilege of knowing him.

Osborn. Prof. Osborn, as we knew him, was an expert teacher of entomology, even in temperament, always ready to help a pupil who was really seeking information. Every one liked to go to his class and witness the ease and perfection with which he could draw insects, apparantly with one hand as good as the other. He was also credited with the very rare accomplishment of being able to draw equally as well with both hands at the same time.

I never knew him to get thoroughly out of patience but once, and that was with a member of the Sophomore class, with whom he had labored long and hard to enthuse with the beauty and usefulness of his subject.

The question was to describe the nervous system of a beetle. The young fellow proceeded thus:

"The nervous system of a beattle begins with a number of ganglion on either side of the thorax and extends entirely down on either side of the backbone." You will agree with me, I am sure, that his disgust was perfectly justifiable. Many are the favors and kindinesses, I received at the hands of this good man.

Miss Roberts was a teacher of rare ability. Her chief delight seemed to be that of helping the backward student. And many, many are the men and women today who rise up and call her blessed, for the help she gave them in more ways than one. I take especial delight in registering as one of that number.

Prof. S. W. Byer was also one of my teachers whom I loved very dearly. The thing that possibly impressed me most was, his youthfulness, superior ability as a teacher in the broadest sense of the term. I longed to be, and wondered if I could ever be like him.

Prof. Marston was another very remarkable man, to whom I owe much of my success; He was stern, exacting and with all this, the best of all was, that whatever mark he gave you, you may rest assured that you earned it. The success of the students in his particular line are the highest testimonials as to his ability as a teacher.

Prof. Curtis. Every young man wanted to be like Prof. Curtis when they grew up, if they could; but the standard of excellence was set so high, it seemed well nigh impossible. Prof. Curtis will never know in this world how much he inspired and helped me as a student; for which I am more thankful than any words of mine can express.

Prof. D. A. Kent. Unfortunately for me, I only had Prof. Kent as a teacher for a short time. I would like to see the student with any brains who did not love Prof. Kent, indeed he was not only a good teacher but fatherly and made you feel so perfectly at home in his presence.

Prof. H. Knapp, as we delighted to call him was a man of rare accomplishments and worth to every student. His advise was so wholesome, sound and I never in my entire six years at the College knew him to turn a student or anybody else away, without giving them good wholesome and inspiring advice when they needed it. I am now reaping rewards from advice he gave me when I first came to the Institution.

Prof. H. Wallace is now Hon. Secretary of Agriculture, Washington, D. C. The heights to which he has risen testifies more strongly than any words of mine can. No one missed Prof. Wallace's class, if they could help it. He was a born teacher; a man too big in heart, mind and soul to be little in any particular. He, like all of my teachers will never know how much he enthused and inspired me, as a student.

Dr. Stalker. While I did not come directly under Dr. Stalker, as a pupil, I would put myself in his way every chance I got. I never found him to busy or too tired to talk with and advise me. He was a man of rare attainments; artistic in temperament; a great student of nature, and a man whose very presence was inspiring and uplifting.

Prof. Bennett gave me my start in Chemistry for which I am very thankful. I wish he knew how much he helped a poor struggling colored boy. I am sure the great and good God will reward him accordingly.

Miss Doolittle. To Miss Margarette Doolittle, one of the best teachers I have ever met, I owe a lasting debt of gratitude. No student missed Miss Doolittle's class unless compelled to do so.

Miss Stacey. To Miss Stacey, I owe not only my knowledge of music; but many other useful things to round out my life, which shall always be a cherished and graceful legacy to me.

Prof. Patrick was my teacher of Agricultural Chemistry. His memory is loved and revered by me. He seemed to take special delight in giving me information, both in and out of the class room.

Geo. W. Carver

By the early 1920s, as Carver and his experiments became more widely known, he began to receive many awards for his work. But one honor, an honorary doctorate from his alma mater, eluded him. In April 1923 he wrote to Pammel asking about the possibility of obtaining the honorary degree. The letter carried the notation "Confidential" written across the top.

4-24-23[19]

My dear Dr. Pammell:—

For some months you have been on my mind more than ever. . . .

My life has been crowded unusually full for some months, being out on lecture tours almost continuously since the 6th of Dec.

The papers have been very liberal and even profuse in their "write ups." The Atlanta papers, both the Constitution and the Journal carried elaborate "write ups" and my picture, I am inclosing one from the magazine section of the Journal.

The St. Louis Globe Democrat carried a fine article and a much finer editorial comment.

May issue of Popular Science Monthly has a picture and an interesting write up.

There is some talk (in the air only), of having one of our large colleges confer a Dr.'s degree upon me.

I do not covet this much from any "Institution" except Ames (I.S.C.).

You giving me the start you did while there and continuing on through all the years I have been out, I have just been wondering if my institution would not like to honor me and my race in this way.

Of course if I am not worthy of it I do not want them to do it.

Now as far as I individually am concerned I am not so much interested but it would help and encourage my race greatly.

Confidentially Dr. Pammel, what do you think of it.

Please tell me frankly, I have always taken your advice, and always intend to. . . .

> Sincerely yours,
> Geo. W. Carver.

Not surprisingly, Pammel supported the idea, and he wrote Carver to tell him so. Carver responded with this letter.

5-5-23[20]

My dear Dr. Pammel:—

Thank you so much for your letter and inclosures. This is just like you for the world. I did not mean that you should do this. I simply wanted the information that I might qualify if I could.

You may be surprised to know that I wore the suit you fooled me into the store and bought for me while in school. I wore it to a banquet, suit, hat and gloves, the people thought it new. Kindest regards to Mrs. Pammel.

> Sincerely yours,
> Geo. W. Carver

But the honorary degree was not to be granted; in fact, Pammel's efforts on Carver's behalf caused Pammel unspecified problems. In July of the same year, Carver penned this note to his mentor.

7-6-'23[21]

My dear Dr. Pammel:—

Thank you for your good letter of some days ago. I certainly thank you for your efforts to secure the degree, I am sorry that it caused you so much trouble.

I do not care for it especially, from any other institution, I certainly will not seek it; if thrust upon me like the Spingarn medal I will take it of course.

I wanted it as a priceless heritage from your Dept. As I believe it inspires more boys and girls to delve into the mysteries of nature and nature's God than any other. . . .

Sincerely yours,
Geo. W. Carver

P.S. This medal is given annually for the most distinguished achievement by an American citizen of African decent.

Carver never did earn a doctorate, although he, was addressed as "Doctor" by most persons for the bulk of his career. In 1926 he explained to Pammel how he had arrived at that title.

Nov. 26-26.[22]

My dear Dr. Pammel:—

The prefix "Dr." as attached to my name is a misnomer. I have no such degree.

It was started fully 25 years ago, by a Mr. Daniel Smith, expert accountant, who came down year by year from New York City to audit our accounts. Sent by our Board of Trustees.

He was greatly interested in my work, and said have you a Dr's degree, I said no. Well he said you ought to have it, your work really more than entitles you to it.

So from that time on he called me Dr. others took up the refrain, he put it in the news-paper articles he wrote. I was powerless to stop it.

I regret that such an appendage was tacked on but I cannot help it.

Yesterday, I thanked God again and again for the life of my beloved teacher and friend Dr. Pammel.

I am constantly praying that God will ever bless keep and guide you. . . .

Most greatfully yours.
George W. Carver.

Pammel retired in 1924, and Carver wrote the following letter of appreciation and reflection when he heard the news.

February 22, 1924[23]

My dear Dr. Pammel:

It is my understanding that you contemplate retiring from active service as Professor of Botany at Iowa State College. In this action, there comes to me a mingled feeling of joy and sorrow.

Joy, because even though you officially retire, you will not wholly separate yourself from the Institution. Sorrow, because the many boys, girls, men and women, not only from Iowa and contiguous States; but all over the country, will sorely miss the love and inspiration as well as superior instruction from you, the Prince of Teachers. Just to walk through the campus occasionally, and let the students see the man who has meant so much to the State, the College, yea, the whole country, will enthuse and inspire many to do their very best.

Personally, I have no words at my command that adequately express my gratefulness to you for the very personal interest you took in me, not only as a student, but it has followed me all through my career, and whatever success I have been able to attain is due, in a very large measure, to you, my beloved teacher, christian gentleman, and friend.

I pray that God's richest blessings may continue to lead and guide you in whatever the future has in store for you.

Your very grateful pupil,
Geo. W. Carver

Pammel replied to Carver, praising his former student and telling him about a testimonial dinner held on Pammel's behalf at which a letter of praise from Carver had been read. Carver responded:

6-8-24[24]

My beloved teacher Dr. Pammel:—

Your remarkable letter has been here several days. I have read it over again and again many times. The more I read it and study it the more I see of the real Dr. Pammel shining through, the thing that above all other things makes you the great teacher that you are. Your strong humanitarian impulses and greatness of soul has always shone through and dominated your life.

I am confident my life and career could never have been what it is had it not touched yours. I wish so much I could have been at that Cosmopolitan dinner, I would have taken great pleasure in telling them to whom I owed a deeper debt of gratitude for my success than any one else, to you Dr. Pammel. . . .

I certainly have done nothing for you to thank me for, the letter to which you refer was very hastily and very poorly done. I have no words to express my appreciation. I considered it a great privilege when I was asked.

Most greatfully yours,
Geo. W. Carver

In 1926, an article celebrating Pammel's work appeared in an Iowa State publication. After reading the piece, Carver wrote to Pammel, adding his own praise, and continuing to rely on his old mentor's scientific expertise by asking him a question.

9-8-26[25]

My beloved teacher, Dr. Pammel:—

I made no effort to keep the tears of joy out of my eyes, when I saw your fine picture, and read the beautiful tribute to your great work in the Aug. Alumnus.

Hundreds of students, North, South, East and west who have had the rare privilege of coming in touch with you in whatever capacity it happened to be will rise up and call you blessed for such contact.

Dr. Pammel is it possible for the sex characters in trees to become facultative? or in other words have the power of changing from one sex to another?

There are three cedar trees in my yard, one has always been pislitate, and the other two bore only staminate flowers until this year when one of them is loaded with berries, and the other one has two berries on it.

They are very old trees and are in close proximity to each other.

Much love to Mrs. Pammel.

Very sincerely and greatfully,
Geo. W. Carver

It was Carver's custom to become intimately involved with the families of those whose friendship he valued. He knew the entire Pammel family well, and even after the professor's death remained in close touch with his widow and daughters. In 1931, two of the Pammel daughters visited Carver at Tuskegee. Violet Pammel wrote to thank him for his hospitality, and Carver penned this response.

Nov. 9-31.[26]

My dear Miss Pammel:—

How I do treasure your splendid letter. Just want you to know that no visitors have come to us this year we appreciated more.

To me it was a real benediction, to have you and your sister actually here. To me it seems like one of those delightful dreams we have occasionally.

The time passed altogether too quickly for me.

I am so glad that your dear mother and my beloved friend is comfortably situated.

To be sure she misses Ames and most of all her sainted husband, O how I love him, marvelous man he was, the likes upon whom I shall not soon look again, if ever.

Just remember, that you and all of the family will always find the latch string of the door on the outside for you, whenever you can find it convenient to come.

Personally I consider it a great honor to have had you and your sister with us, so you could see some of the fruits of your dear Father and Mothers labors.

I am greatfully yours,
G. W. Carver

When word reached Carver of Mrs. Pammel's death in 1935, he wrote a letter of consolation to Violet and tried to summarize what her mother and father had meant to him.

3-17-35[27]

My esteemed friend, Miss V. Pammel:—

Your fine letter has just reached me and brought so much comfort.

I saw in the last issue of the Alumnus, that your dear mother had entered into the fullness of joy. I was greatly shocked as this was my first knowledge of it.

I am sure all of you children can appreciate my feeling, as your home at Ames was my home, and your Sainted Mother and Father never tired of doing lovely things for me, not only while I was in school, but throughout the years which followed.

When your mother was here she impressed me as being more angel than human, and I shall always treasure her visit here, as well as Dr. Pammel's, one one of the dearest experiences of my life down here.

I do not need to tell you how much I enjoyed you and Mrs. Seale.

I want the rest of the family to know that I share with you this great personal loss.

Their memory to me is sacred, and makes Heaven so much nearer and dearer to me.

Another former Iowa Agricultural College professor who Carver stayed in touch with was James Wilson, who left the college in 1897 to accept an appointment as U.S. Secretary of Agriculture under President William McKinley. Wilson continued in that position until 1913. A 1904 letter from Carver to Wilson was signed by Carver "Your pupil":

July 23, 1904[28]

My dear Prof. Wilson,

I feel that it is now time I received a letter from you. I am none the less your pupil now than when I was at Ames. I am just as much dependent upon your council and advice.

I follow you pretty closely through the papers and personal friends, but a letter from you brings me so much nearer in personal touch with you which of itself is indispensable.

I am now at Knoxville Tenn. Delivering a course of lectures at their summer school, on nature study and Agr. The advance of magnolias set for you are doing nicely.

An Agrl. Bldg. to cost near $1,400 is in the course of erection and I am very proud of it.

Dr. Grannahan wishes to be rememberd to you. My work at Tuskegee seems to be moving along fairly well. My new cotton is shewing up wonderfully well this year.

Both corn and cotton crops are above the average in Macon Co. (Ala.) due we think largely to the influence of the lessons gained from the school.

We have a farmer's institute which meets every month, last tuesday more than 100 people were present.

I have a fine opportunity to go to Porto Rico at a good salary as Agnt. Teacher. What do you think of the possibillites in this direction of those foreign islands.

Very sincerely yours,
Geo. W. Carver
Your pupil

Another of Carver's friends from Iowa State was Henry C. Wallace, an agriculture professor who later became Pres. Warren G. Harding's secretary of agriculture. Wallace's son, Henry A. Wallace, whom Carver took on nature hikes while at Iowa State, succeeded his father as secretary of agriculture and later served as vice-president under Franklin Delano Roosevelt. The younger Wallace remembered Carver as the "kindliest, most patient teacher I ever knew."[29] This letter of 6 January 1924 gives some indication of the esteem in which Carver held Professor Wallace.

1-6-24[30]

My dear teacher and friend:-

I believe more firmly in the (science?) of telepathy. I had been thinking of you more than usual here of late and "How God" had called you to perform the great service to humanity that you are now rendering.

I pray that He may continue to give you the guiding light that has attended your administration up to date in such a pronounced way.

Of course I never can repay you for being so kind, and indulgent to a poor little wayward black boy when in school.

I wish that God in some way would shew you how I appreciate it, and reward you accordingly. . . .

Geo. W. Carver

Carver knew that the debt of gratitude he owed his friends from the pre-Tuskegee days was great. They had invested time, money, and hope in him, a "poor struggling colored boy." It was important to him to let them know that they had invested wisely. The Iowa memories, in particular, were sweet. As the years passed and his difficulties at Tuskegee increased, those memories grew even sweeter.

FOUR

Tuskegee Institute

Carver and His Coworkers

I think I shall not be here much longer. . . . The school has not kept its promises with me and I do not think will. . . . I think I have parleyed with them about long enough.

Geo. W. Carver 2 July 1912

BOOKER T. WASHINGTON ran Tuskegee Institute with the strictness and regimentation of the most efficiently operated Old South plantation. He expected total commitment and even obedience from his subordinates. He measured their worth by the degree to which they were willing to overtax and inconvenience themselves for the Tuskegee cause.

Washington was also a practical man. He saw no value in research that failed to produce near-immediate results. Visionaries and idealists annoyed him; he wanted to be able to tally the day's productivity at sunset each night.

George Washington Carver, by contrast, was reluctant to be inconvenienced by anyone. His sense of having been chosen by God to do wondrous work, combined with an educational background equaled by few blacks, either at Tuskegee or in the nation at large, left him feeling entitled and impatient with those aspects of his job that seemed tangential to what he saw as his real purpose. While Washington expected him to suffer the same privations imposed upon other faculty members, Carver continuously sought additional resources and freedom from mundane administrative duties so that he could devote his energies to his ostensibly divinely appointed calling. His fellow faculty members and Tuskegee's administrators frequently interpreted this as a desire for preferential treatment. When Washington asked him to become more productive and to complain less, the middle-aged scientist complained even more about the work already expected of him.

It is probably true that Carver had no intellectual or creative peer at Tuskegee, but he acted as though he knew it, which created tension between him and his coworkers. Not only did he engage in running, and often petty, battles with Principal Washington, but he regularly found himself embroiled in disputes with his colleagues. Not surprisingly, many of them resented him deeply.

Many others, however, admired Carver just as deeply. His intensity and sincerity, his capacity for love and friendship, captivated many persons at Tuskegee who became his life-long friends. Some of those friends, such as Mr. and Mrs. Harry O. Abbott, like Carver himself, never really forgave his Tuskegee critics.

During Carver's last year as a student at Ames, Iowa, Booker T. Washington heard about the eccentric young black scientist who was making a name for himself at a Northern white school. Washington wanted to attract Carver, then in his early thirties, to Tuskegee as head of his about-to-be-established agricultural department. He wrote to ask Carver if he was interested in the job. Carver responded with a letter that sent an ambivalent message: the body of the letter said he was not interested but a postscript implied that he was.

4-3-96[1]

Dear Sir:-

I have just returned from a lecture tour and found your letter awaiting me.

I will finish my masters degree in scientific agr. this fall, and until then I hardly think I desire to make a change, although I expect to take up work amongst my people and have known of and appreciate the great work you are doing. Mississippi has been negotiating with me for some time (Alcorn A. & M.). I was ready to go this spring, but a long line of exp. work has been planned of which I will have charge. And from the educational point of view I desire to remain here until fall.

Very respectfully

Geo. W. Carver

P.S. Should you think further upon this matter I can furnish you with all the recommendations you will care to look over.

Carver was obviously eager to impress his potential employer, and also to convince the Tuskegee principal that he did have other options. But apparently he was afraid that he might have overdone it. Two days later he penned another letter to Washington, suggesting that he was not as intransigent as his first letter had indicated.

4-5-96[2]

Dear Sir,

I sent you a few days ago an answer to your inquirey, being away occasioned the delay.

Of course I should prefer to stay here until fall as then I will receive my master's degree, but should I get a satisfactory position I might be induced to leave before.

I can send an abundance of recommendations if needed.

Very Resp.

Geo. W. Carver

A few days later, Washington again wrote, expressing interest in Carver. Carver responded with a much longer letter, again letting Washington know that he did have other options and that he was committed to the cause of black education.

April 12, 1896[3]

My Dear Mr. Washington:

Yours of April 1 just received, and after a careful consideration of its contents. I now venture a reply. It is certainly very kind of you to take the interest you have in me.

Of course it has always been the one great ideal of my life to be of the greatest good to the greatest number of "my people" possible and to this end I have been preparing myself for these many years; feeling as I do that this line of education is the key to unlock the golden door of freedom to our people.

Please send me catalogues and any other data you may have with reference to your institution, so I may get some idea of the present scope of your work and its possible and probable extension. I should consider it a very great privilege to have an interview with you, but cannot say if I will be in the west or no. As among the prospective locations, I accepted a position within the shadow almost of your own institution, and nothing more remained to be done but the election to the chair, but said election was deferred until spring, and will take place very soon now. So if you are prepared to make me an offer now it shall receive my first consideration. . . .

Should I not accept the position above mentioned I will be here at the college all summer except when my occupation calls me away. At the next writing I hope to give you a more definite answer.

May the Lord bless you and prosper your work.

Geo. W. Carver

On 17 April 1896, Washington offered Carver one thousand dollars a year, a substantial salary at the time, plus room and board, to join the Tuskegee faculty. Committed to the notion of maintaining an all-black staff at his school, Washington told Carver he was the only "colored man" in the country qualified to head the newly created Agricultural School. But Washington also wrote that he wanted people at Tuskegee who came to help the race, not just for the money. Carver responded immediately.

April 21, 1896[4]

Yours of the 17 inst., just received and contents carefully noted. I hasten to reply because I must decide very soon whether I am to stay here, go somewhere else in the South, or come to you.

I am very much pleased with the spirit of your letter, and assure you if I come the money will not be the sole object, only secondary. One institution near you offers me the same as you with the understanding that my work is to recommend for me an advance in wages also a house. I already have a position here as you will see by the letter head and one of my professors told me today they would raise my wages here if I would stay, but I expect, as I have already stated to go to my people and I have been looking for some time at Tuskegee with favor. So the financial feature is at present satisfactory. I shall be pleased to do all I can to cooperate with you in the broading of the work and making it far reaching in extent.

Enclosed please find the 4 years course of Agr. from which I graduated in '94. I had special training in the ones I have marked aside from the regular course.

I would like to know if your course in Agr., will be something of the kind, and what I will be expected to teach.

I have been for several years making large collections along economic lines and should I come to you I will bring my collections and cabinets with me for use.

When would it be necessary for me to come? I will finish my P.G. work this fall.

If the course of study is not something out of my range I will accept the offer. I shall await an answer with a considerable degree of anxiety as I must decide soon.

May God bless you and your work,

Sincerely yours,
G. W. Carver

Washington reassured Carver that the latter's training made him the best-qualified black person in the country for the job and said that Carver could postpone coming to Tuskegee until he had finished his graduate work. Carver responded eagerly and also indicated his support of Washington's "solution to the 'race problem.'"

May 16, 1896[5]

My dear Sir:

I am just in receipt of yours of the 13th inst., and hasten to reply.

I am looking forward to a very busy, pleasant and profitable time at your college and shall be glad to cooperate with you in doing all I can through Christ who strengtheneth me to better the condition of our people. Some months ago I read your stirring address delivered at Chicago and I said amen to all you said, furthermore you have the correct solution to the "race problem."

I learn that you are to be within a short distance of our college this summer at a chautaqua, convention or something of that sort.

Now I would be delighted to have you stop and see me, also to look our college over, please do so if you possibly can. I will be pleased to meet you at some point near if you cannot stop at the college.

It is very kind of you to keep me posted as to the progress of the work and to allow me such ample time for preparation.

I presume you overlooked the little slips you spoke of inserting.

Providence permitting I will be there in Nov.

God bless you and your work.

Geo. W. Carver

Linda McMurry has written that Carver arrived on the Tuskegee campus feeling "invincible."[6] He expected to be treated deferentially, only to discover that the more he asked for, the more hostile his coworkers became. He wanted two rooms to live in, for example, at a time when campus bachelors were expected to live with someone else in one. Within a couple of months of his arrival at Tuskegee, frustrated in his efforts to obtain the perquisites he expected for a man of his station, he appealed directly to the school's finance committee. That letter created an irreparable strain between himself and much of the Tuskegee faculty.

November 27, 1896[7]

Messrs of the Finance Committee:-

Some of you saw the other day something of the valuable nature of one of my collections. I have others of equal value, and along Agr. lines.

You doubtless know that I came here solely for the benefit of my people, no other motive in view. Moreover I do not expect to teach many years, but will quit as soon as I can trust my work to others, and engage in my brush work, which will be of great honor to our people showing to what we may attain, along, science, history, literature and art.

At present I have no rooms even to unpack my goods, I beg of you to give me these, and suitable ones also, not for my sake alone but for the sake of education. At the present the room is full of mice and they are into my boxes doing me much damage I fear.

While I am with you please fix me so I may be of as much service to you as possible.

Also I am handicapped in my work, I wanted a Medical Journal the other day in order that I might prescribe for a sick animal. It was of course boxed up, couldn't get it. Trusting you see clearly my situation, and will act as soon as possible, I remain most

Resp. yours

Geo. W. Carver

Telling his coworkers that he intended to stay at Tuskegee only a short time, until he could return to his "brush work," impressed no one. In fact, it did just the opposite. Why, many of his fellow workers must have thought, should they give this prima donna anything if he only intended to leave eventually?

Over the next couple of years, Carver's frustration increased. He felt that he could accomplish a great deal if he were supported; instead, he considered himself surrounded by incompetent subordinates, overworked by the Washington administration, and poorly supplied. In May 1898 he wrote to Washington in an effort to assess the situation and to suggest that if something were not done to improve his working conditions, he would consider leaving Tuskegee.

5-30-98[8]

Dear Sir:-

In making this report, I think it but fair to you, fair to the school and fair to myself to make this statement of facts.

I assure you that no one is more deeply interested in the welfare of the school than myself, and especially my Dept. I have labored early and late and at times beyond my physical strength; have not asked for a private secy. and if you look into my work carefully you will see that I need one quite as much as some who have two.

I had made partial arrangements to enter the Shaw School of Botany, St. Louis, from which I hope to take my doctors' degree, a degree that no colored man has ever taken, but your many letters urging the cutting down of expenses, and your desire to have me study the food question, in which I am also deeply interested, and the very important relationship the farm as a whole stands to the financial side of the school I canceled my engagement with Shaw.

Again—In taking charge of the Agr. Dept. it was my understanding that you wanted it to grow. I have put forth every effort in that direction that time, means and opportunity would permit.

I have been looking forward to a Dept. second to none in the U.S. in the matters of equipments, methods of teaching and results obtained.

We have great fruit possibillities here, and instead of spending so much money every year in purchasing trees, vines and shrubery, of various sorts I had planned to do our own budding, layering, grafting and inarching ourselves, cutting down that expense, and at the same time have fit subjects for the students training a thing of beauty for visitors both north and south. The landscape feature of our gardens and grounds I expected to have second to none.

Farm

On the farm I expected the same improvements incident to it, viz. a moddel in every peticular, bending all our energies toward the saving of food on one hand and the production on the other, hoping under favorable conditions of the weather to cut down the hay bill two or three hundred dollars. Mr. Menafee will work in connection with the farm along this line.

Barn

Here we have just decided, and quite wisely, to increase the number of hogs, I am looking forward to the time when we will furnish nearly or quite all of our beef mutton and pork. To do this we must make a steady growth.

To accomodate the number of brood sows we expect to have it will take about 40 new brood pens, which must be built and 3 more open pens, to separate the different hogs, sows, pigs etc. as necessity demands.

In fact I was going to ask you to reccommend to Mr. Chambliss that he must help along that line or be decreased in salary, but not excused altogether, as he is a good man with the cows. If we are going to get the number of hogs we spoke of the above mentioned work must be done, or it will be that much time and expense practically lost.

Besides the dairy herd will always be just as it always has been as long as we bend all of our encrgies to build it up 9 mo. and demoralize it 3. We have now under Mr. Chambliss nearly 60 cows 30 calvs and the cull and I had planned to bring all of our dairy stock from Marshall farm as they do so much better here, besides they are not kept track of down at Marshall, and some of them die outright, others are sent up here for beef, and soon we are spending a lot of money for five cows to repeat the same process, 100 head of stock will if propperly cared for keep any one man busy.

The records for last year are about as follows for the 3 mo. vacation, accidents due to irresponsible students, 1 colt value $60 2 cows one worth 75 and the other 80 = $165 saying nothing about the multiplicity of pigs, and a number of calves.

Here I am working with the smallest and most inexperienced staff of any station in the U.S. having the fourfold object in view that we have, also with less means to cary on my work. It is impossible for me to do this work without men and means.

Now Mr. Washington, I think it ludicrously unfair to have persons sit in an office and dictate what I have to do and how I can do it, also to tell me what has been done, If I thought things were to run as they have always run I would not stay here any longer than I could get away. I did not come down here to make experiments to find out what could be done with our Southern soils. I know what is needed.

You know they said clover would not grow here, scoffed at the sweet potato exp. while they were growing and some even now want to deny the yield because they would not go to see it, (they are among the teachers to). It is the largest yield ever made in the state. The acorns were brought under rigid protest it was even accused to my face (by a teacher in high authority.) as going crazy, yet Auburn in making up its recent list of valuable hog food includes the acorn. Miss says "it is one of the most important subjects taken up by any station."

Here are the exact words of my college paper of last mo. "A short time ago the student gave a biographical sketch of our illustrious alumnus, Mr. G. W. Carver '94. A bulletin from the Tuskegee Experiment Station has just reached our desk. This station has but recently been established particularly for the benefit of the colored race of the state and Mr. Carver has been appointed director. His first bulletin treats of the value of the acorn as stock food. He lucidly demonstrates the practical economy of a new instruct along this neglected line. Judgeing from the features of this bulletin, Mr. Carver surely is and will be a powerful factor in the development of the 'New South.'"

Mr. Washington I simply want a chance to do what I know can be done, not what I think.

The records from Marshall Farm for the period of 3 years show that 30 heifer calves have been sent to the Farm and only 5 have ever been returned to the dairy.

These were our best calves.

Mr. Holland is given 3 men 4 if needed to tend 10 acres. Mr. Green 10 men to tend 100 acres I have 20 acres in Exp. work and not a man given me. I want to conduct some feeding exp. and of course must have responsible to do them.

I have no further recommendations to make. A number of our finest cows are to drop calvs soon, and during this hot dry weather skill and vigilance must be exercised to save them. I say all this in behalf of my Dept. and the best interests of the school. Respy

Geo. W. Carver.

Despite Carver's plea, and the legitimacies of many of his complaints, no substantial changes were made in his workload or in Washington's expectations of him. Carver kept writing to Washington, asking for more help and cooperation in running his department. Again, it seemed inconceivable to him that someone of his stature and training should be denied anything, as he made clear in this letter to the principal.

1-17-02[9]

Mr. B. T. Washington,

From causes which I am wholy unable to decipher, I feel I do not get the cooperation of the council, many times no attention is paid to my wishes and things passed over my head which work contrariwise to my efforts to cary out the schools wishes, as in the case of last night.

Yesterday I spoke very plainly to Mr. Owens & Mr. Crawford about having old delivery and things that did not belong to the barns and dairy lyeing around. This morning I must tell Mr. Owens to put these back and let others come in, you and others go over and criticize them, as you ought.

It would have been much better to have had Mr. J. H. Washington give up our quarters at the old barn, which I claim he has no right to rather than cumpel us at the dairy to do things which we say in the classroom to the student is wrong. I shall endeavor to cary out the school's wishes to the letter, but I do not like this decision. My work has been hamperd and renderd unsatisfactory the entire school year from similar reasons.

I write in this way as I feel that you will care as they are interfering with the efficiency of my work.

I want it understood that this is in no way rebellious, but a plan statement of facts as I see them.

Yours very truly,
Geo. W. Carver.

Carver found the entire atmosphere at Tuskegee debilitating. Much of the problem, he wrote Washington, could be traced to the low standards established by the faculty. He urged the principal to take corrective action.

Sept. 13, 1902[10]

Mr. B. T. Washington:

Respecting your note about suggestions to bring before the teachers at the 11 o'clock meeting today, it occurs to me that it will be well to speak to the Heads of Divisions about proper co-operation. Some of them—in fact, altogether too many, have simply the one idea in mind,—that is to look after their own particular divisions, which, of course, is correct to a degree, yet the broader and more liberal view should be taken;—that is, each division is only a fraction of the great whole and that each work to the interest of the school as a whole. Wherever one division can accomodate another, it should be done. This, in too many instances, is not done now.

I also think it would be well to call attention to the use of slang by the teachers and especially in the presence of students. I have noticed quite a little bit of this already. The matter of teachers calling each other by their given names—such as, "Hetty," "John," Bill," etc., should be corrected. Also, their addressing each other thus: "Hello! How are You?" etc.

I dare say that those in charge are doing all they can to improve their methods of teaching, but if you think it wise, it might be well to call attention to the value of the objective method of teaching and the most excellent facilities we have for that kind of teaching; and that no teacher should go to the class room to talk abstractedly. Almost every line of teaching can be demonstrated in a greater or less degree from actual material to be had here, upon the ground. I think it well also to speak of the high standards as I think the major part of our trouble with both students and teachers comes from the standards being too low. I am sure this is the main cause of our trouble in the agricultural department. Nothing but the highest standard can bring the success which should come to an institution of this kind.

I presume you have in mind to speak to them about the personal interest which should be taken in each student—especially in students who seem rather dull and are from any cause behind in their studies, deportment, etc. I refer to studies mainly but do not undervalue deportment, etc. These are merely suggestions which you may have in mind already.

Yours truly,
Director of Agriculture.

There can be no question that Carver's working conditions were challenging. Booker T. Washington was a micromanager who was often gone from the campus on goodwill and fundraising trips. In his absence, Washington usually left his half-brother, John Washington, in charge. Carver's relationship with John H. Washington was even more strained than was Carver's with Booker T.

When Booker T. was on campus, Carver's co-workers often complained about him and his work to the principal, without first talking to Carver or giving him a chance to address the issues. This practice bothered Carver greatly. In September 1897 he wrote to Washington, urging him to put a stop to this practice: "Hereafter if persons will make complaints directly to me . . . I will do the best I can."[11] Some months later, with the practice still going on, Carver wrote to Washington, telling him, "I wish you could be here more than you are and look into much yourself and not take people's word for it."[12]

But Washington himself could not be relied upon to be reasonable in his criticism of Carver. On one occasion in 1899, Washington wrote to Carver, pointing out to him that his half-brother had complained that Carver was late to a meeting of "Directors and Division Instructors." Washington admonished, "It is important that these meetings be attended promptly." Carver responded tersely, pointing out to the principal that he was late because "I was detained by you in your office."[13]

The most bitter, and ultimately most demeaning, controversy Carver became engaged in at Tuskegee was his battle with George R. Bridgeforth, a young man who came to Tuskegee in 1902. Bridgeforth, described by Linda McMurry as Carver's "temperamental opposite," quickly developed disdain for Carver's incapacity for administration and won the ear of Washington.[14] He began writing long letters, both to Washington and Carver, criticizing the latter's handling of the agricultural department. The gauntlet had been thrown down. The sensitive and prideful Carver could not allow such impudence to pass unnoticed.

The battle between Bridgeforth and Carver came to a head in 1904. Several issues divided the two men, but the most pronounced was Carver's handling of the institute's poultry. By all accounts, Carver failed dismally in overseeing poultry operations at Tuskegee. In September 1904 a special committee investigated conditions and reported that the school's poultry flock was in very bad shape and even suggested that Carver had falsified his reports to Washington so as to cover up his own managerial ineptitude. Carver responded with the following letter.

Oct. 14-'04[15]

My dear Mr. Washington,

I beg to acknowledge receipt of your note, and come now to the most painful experience of my life.

For seven years I have labored with you: have built up one of the best Agl. laboratories in the south, so much so that the people of your own town recognize its value.

Only yesterday the Tuskegee Cotton Oil Co. submitted samples of cotton seed meal, and cake for analysis and have arranged to bring in samples every week.

The museum is the best of its kind in the south and constantly growing. The Experiment Station, in the nature of the problems chosen and the results obtained I am sure has no equal south. Now to be branded as a liar and party to such hellish deception it is more than I can bear, and if your committee feel that I have willfully lied or party to such lies as were told my resignation is at your disposal.

I deeply regret to take this step but it seems to me the only manly thing to do.

I shall always feel kindly to your work and shall continue to be loyal to Tuskegee and its interests.

Yours Very truly,
Geo. W. Carver.

Bridgeforth knew just when to strike. He encouraged Washington to relieve Carver of some of the latter's responsibilities. Washington appointed another committee to investigate Bridgeforth's recommendation. That committee concluded that agricultural responsibilities should be divided between Carver and Bridgeforth, with the former as director of the experiment station and agricultural instruction, and the latter as director of agricultural industries. It was the committee's effort to separate theory from practice and assign each man the tasks for which it thought he was best suited.

In early November, Washington sent Carver a copy of the committee's report and asked for his response. Despite the fact that Carver had often complained of being overworked, he rejected the committee's recommendations in this letter to Washington. He pointed out, also, the patent falsehood of many of the allegations against him, and, once again, offered his resignation.

Nov. 8-'04[16]

My dear Mr. Washington:-

I beg to acknowledge receipt of your communication of Nov. 5 enclosing the Committee's recommendations for reorganization, and make note of the following things, to which I do not see my way clear to accept for the following reasons.

(a) I do not agree with the title, it is too far a drop downward. A few at Tuskegee will understand it but the public never.

(b) The 7th is also very embarrassing for days I have been without a stenographer and Committees pressing me besides the other work.

(c) The 8th and most important of all is the split in the teaching which could not help but result in constant turmoil and final failure. I beg of you Mr. Washington for the sake of your Agr. to maintain your organization, having a Dept. head with logical subdivisions.

Your committees report if you note, has no Dept. but a flimsy organization which cannot stand, and will furnish data for constant outside criticism. No such arrangement exists in any other Dept. of the whole school.

The above and much more which I will tell you if you grant me the privilege of a hearing, causes me to ask that you kindly accept my resignation to take effect just as soon as I can get the herbarium and cabinets labled and in place where they will be of the highest service to the school.

I am going over my private collection now laying out quite a few things which I shall not care to take with me and which will add greatly to the museum. Kindly permit me to make this request that you may have it kept intact the exclusive property of the Agrl. Dept. and one of the monuments to my seven and one half years labor at Tuskegee.

As soon as my mind can get sufficiently clear I shall finish the manuscript for the Exp. Station Rept. and turn the books over in good shape so the work need not suffer.

I cannot refrain from calling attention to these grevious and damaging errors in the committee's report on inspection.

"No native birds appear in the exhibit."

In a casual way I counted 40 this morning that are native to our state.

"Mr. Carver claims that he has done all of the work in the Experiment Station fields even to the extent of dropping the seed in the various beds."

I make and made no such claim. Another point of dissatisfaction is that in your present arrangement I have absolutely no connection with the farm. To this I could not consent. I regard it as very unfair to take the poultry yard from me at this time and emphasize the peculiar conditions existing there in a most embarrassing way to me.

I had a talk with Mrs. Washington, we saw it alike, and agreed to help me all she could. I put in $25 of my own money because I had made up my mind that the thing should not fail. I considered it the manly thing to do.

Another thing which weighed and is yet weighing heavily upon me is this, your committee composed of very thoughtful persons forgot that I had spent nearly 7 1/2 years of faithful service at Tuskegee; it came a day later apparantly as an after thought.

Pardon me for writing such a long communication. This is in no way intended to cover up or condone my short comings.

I want this clearly understood that from now on until my work is finished my whole heart and spirit shall be put into the work.

With highest esteem for you Mrs. Washington and my other friends.

I am yours with very best wishes
Geo. W. Carver

Carver backed off of this intransigent stand a few days later and wrote to Washington in a more compromising tone.

Nov. 14-'04[17]

My dear Mr. Washington,

In considering your note and the committees report in detail, think if there is to be no Dept. head I should bear, at least an advisory relationship, to it.

The Supt., or Direc. of the farm as you shall call him should with his corpse of assistants make out the policies for the operation of the farm in all of its divisions covering in their order spring, summer & fall. These should be submitted to and approved by the Gen. Supt. of Industries, Advisor, the Principal and Treas. of the school, with whatever changes they may jointly think wise.

Said policies should be placed into the hands of the Gen. Supt. to carry out in detail much as he chooses; under no circumstances should these policies be changed or modified without the approval of all concerned.

As it is now the farm is given up largely to costly experimentation. Should this go into effect I am sure we would avoid the mistake of planting large acreages of Burmuda grass seed.

Buying a milk herd of an advanced age and too far from home to serve the highest good of the school.

The planting of peanuts with their shells on entirely too late for results, hence the failure.

The illy selected land for sugar cane, which meant shortage of crop.

The poorly selected seed for sweet potatoes which has meant a dry, hard inferior lot, both in quality and quantity.

With Mr. Attwell as Supt. I was supposed to have held this advisory relationship but it was very often a farce and the advice ignored altogether at times and especially if a technicallity could be found, due largely to the fact that no clearly defined policy existed. I spent much time in writing out suggestions and recomendation to be ignored afterward.

I was often ignored and am now wholy left out of the meetings pertaining to the farm, and no advice from me is sought except from those who are under me.

My assistant was consulted freely and frankly and sometimes brought into meetings where I should have been, and in every instance he wuld come and ask me what I thought of such and such a thing.

e. g. A large scheme for irrigation was gotten up. Not a word was said to me except as the different members of the committee would come and ask me what I thought of this or that.

A Burmuda grass scheme was gotten up I was ignored in it all. Yet the man who is to put it out comes to consult me as to the probbable cost per acre, whether to plant roots or seed, and asks that I go over and help him select the land.

The Experiment station has practically no relationship to the farm, no advice is sought and but little taken after it is given.

I have yet to learn Mr. Parks' true relation to the school, what he is to do and which not to do, as I was in none of the meetings pertaining to his employment.

The Com.'s Rept.

1st I have no objections to ultimately giving up the poultry yard, but it must be put in good standing by me before this is done if I am to remain in the school's employ. I promised Mr. Baldwin and yourself to raise chickens. I had them there, they got away. I do not know how, (much to my shame and regret.) It is my intention to give you more.

In this connection I would like to ask that Mr. Loring Brown of Smyrna Ga. be asked to come down and give us the benefit of his knowledge in the operation of a successful poultry plant.

He is anxious to do it, and says that in a short time he can train our students to do the same work.

5th I am now teaching 12 periods per week and I feel that this is all that I can do in connection with my other duties.

4th Regarding the course being planned and its inconsistencies I will have to look into these charges. I was ignored by the committee and asked nothing about the course or my teaching.

In conclusion I wish to say that another embarrassing situation is this that whenever I call an Experiment Sta. meeting of the so called Agrl. faculty the people come if they desire and stay away if they like. I have no redress in which I am sustained.

An important matter has been up since yesterday pertaining to a sweet potato house it has been refered to me. I have been unable to get Mr. C. W. Green up to date as the Supt. and other duties are in the way.

I simply cite this as a need of a better organization that will beget cooperation.

The committee suggests that the industrial classes and theory be separated this should not be considered a moment but rather that which will connect them more closely together, so that the theory as taught in the class room and in these classes should be well correlated and consistently carried out in our field operations which is not done now in as large a degree as it should.

I should like to have a stenographer a half day or be given a machine allowed to train a student in to do my work. I think this is due me. I have never asked for a rise in salary, have economized rigidly for the school and I fear to my detriment. An unanswered and unsolicited letter is on my desk now offering $200 more per year besides other tempting advantages. I bring this to your attention for no other purpose than to prove my interest in Tuskegee.

Adenda

Another and more feasible and I think satisfactory scheme would be to split the Dept. into two divisions.

Viz. (a) A training farm in connection with the Exp. station where all of the proffessional agricultural students shall be trained.

(b) A farm for economic production operated mainly by hired labor and night students.

I have not had time to work out details, but if you think it worth considering I shall be glad to do it.

I am sure the farm would come out much better financially and the students be much more pleased and encouraged and be the means of getting many more into the Dept.

In this scheme you could preserve the Dept. heads logically and without embarrassment to any.

Respect. submitted

Geo. W. Carver

There really were no winners in the Carver-Bridgeforth battle, or, if there were, they certainly did not include Carver. To be sure, he gained minor concessions: he retained control over the poultry yard and kept his title. But he had also incurred the wrath of Washington, who, like many a bureaucrat before and since, identified the worth of his managers by their ability to settle problems before they reached his desk.

In 1908 the battle between Bridgeforth and Carver was renewed with even greater intensity, and again the poultry yard served as the battleground. This time the split that had been recommended four years earlier was effected: Carver became director of agricultural instruction and the experiment station, Bridgeforth the director of agricultural industries. Two years later, Washington again took away some of Carver's responsibilities by appointing him director of the department of research and the experiment station only. Neither reorganization satisfied Carver, of course. The fighting raged for several more years with the besieged Washington trying repeatedly to find a solution that would satisfy both of his strong-willed subordinates. Frequently Washington issued ultimatums to Carver; just as frequently, Carver threatened to resign because, as he saw it, promises made to him had gone unfulfilled. In May 1912, he told Washington:

May 4, 1912[18]

Mr. B. T. Washington

I beg to bring the following things to your attention, which I think you will agree thoroughly justifies me in the course taken.

For 16 years I have worked without a single dollar's increase in salary, yet I have sat quietly and seen others go away above me in my own department (at that time.) I have asked only for things to work with and have not succeeded in getting these.

Please read the following all of which, has your signature.

See note of Nov. 23d 1910 in which a number of definite and specific promises were made as follows:-

1st A Department of Research Consulting Chemist, with Prof. Carver as the director of this department, work to embrace the following.

In name the above has been carried out.

1 Experiment Station. Carried out.

2 Bulletin publications, practically impossible to get matter published. If a 50¢ rake is asked for it must go to the budget Committee, which means it will not be gotten at all or delayed beyond the point of usefullness.

3d Bacteriological work, not a single piece of apperatus purchased.

4th Analysis of water, part of the apparatus here.

Milk " " " " " "

Food stuffs for human and stock—part of the apparatus here.

Paints oils ect. part of the apparatus here.

5 Special lectureing on Agriculture and domestic science.

6 Poultry yard.

7 Museum.

2 That a first class laboratory be fitted up for Prof. Carver so as to enable him to cary out whatever investigations he may wish to undertake along the above lines. This has not been done although nearly 18 months have passed.

3d He may teach Agricultural classes if he desires, not true, see note of Jan. 9th 1911.

See note of Nov. 25th 1910.

It is to be understood that Mr. Carver is to have the use of a stenographer at least one half day, and that during this time the machine and stenographer are to be located in Mr. Carver's office. Not true See note of May 4th 1912. in which an arbitrary order (without consultation) for the service of the stenographer just 2 hours per day which shews that that paragraph was not well thought out.

See note of March 2nd 1912. Doubt is expressed as to the wisdom of fitting up the laboratory.

I was told by two Council members to my face that the laboratory would not be established if it were not for me and that if I should drop out it would not be kept up.

All of the above facts and others, together with your note of May 3d. to the effect that my teaching was unsatisfactory, places me in a very embarrassing position.

I interpret the above to mean that the school is realy tired of my services and wishes me to resign.

I see no other alternative, Am I correct.

Very truly yours,
Geo. W. Carver

Carver, of course, never did resign, although he wrote numerous letters of inquiry about other job possibilities. One recipient of such letters was James Wilson, a former Iowa State teacher. Since leaving Iowa, Wilson had served as secretary of agriculture under Presidents McKinley, Roosevelt, and Taft. Wilson, although somewhat perplexed by Tuskegee politics, consistently urged Carver to stay where he was, even telling him that the only way he could enter government service would be to take civil service exams, just like any other job applicant.

Carver, however, refused to see himself as just another candidate for a job. Moreover, his sense of mission in being called by God to serve African Americans in the South mitigated against his leaving Tuskegee. Hence, he found himself inextricably tied to the place that was at once a source of hope and despair.

For Washington's part, his troubled and troubling scientist was still one of his most effective teachers and researchers, and the practical principal knew that Tuskegee could ill afford to lose the person whose work was bringing more praise to his institute than all his other subordinates combined.

What was the effect, then, of the decade-long bickering that had embroiled the entire Tuskegee operation? For one thing, it led Washington to try to move Carver more and more into the area for which he thought the temperamental scientist was best suited: applied research and the supervision of the experiment station. As for Carver himself, Linda McMurry offers this accurate assessment of the Bridgeforth-Carver battle:

> Although the individual disputes were petty, their combined influence on Carver and his work were enormous. Increasingly he retreated from his Tuskegee duties and focused his efforts where they were

appreciated—away from Tuskegee. It is also likely that his administrative failures and humiliations drove him to prove to himself and to others that he did have a special, God-ordained role to play in the destiny of his people.[19]

That was the situation when Booker T. Washington died in November 1915. That Carver was grieved by Washington's death is indisputable. Perhaps he felt no little guilt as well, knowing, as he did, that he had been a source of great anxiety and tribulation to his old friend. Perhaps he simply regretted not having communicated adequately to Washington while the latter was still alive his admiration for the famed educator's accomplishments. Certainly that thought was on his mind when, in February 1916, he wrote to a friend of the principal's, "I am sure Mr. Washington never knew how much I loved him, and the cause for which he gave his life."[20]

But Washington's death had at least one positive consequence for Carver, for it brought to Tuskegee a man who, having come from Hampton Institute, had not been soured by the long imbroglio between Carver, Bridgeforth, and Washington. Robert Russa Moton was an ardent admirer of both Washington and Carver, the latter of whom was approximately fifty years old when Moton took over direction of the institute in 1915. Moton realized what an asset Carver was to the school, and he nurtured Carver's research efforts by giving him what Booker T. had always refused: freedom from teaching duties during the regular academic year.

The Moton-Carver relationship was placid enough. Each seemed to respect the other, and compliments flowed as easily from the professor as from the principal. Unfortunately for Carver, his release from teaching duties, and his increased notoriety among non-Tuskegee admirers, only fueled the fires of resentment and jealousy on campus. This resentment lasted for the remainder of Carver's life, as evidenced by the criticism of Carver that one of his Tuskegee colleagues, Monroe Work, leveled in a review of two biographies of Carver that appeared in the wake of the latter's death. Work felt compelled to remind readers "that [Booker T.] Washington was highly critical of Carver's impracticality."[21]

This is not to suggest, however, that Carver had no friends at Tuskegee, or that his relationships with all of his coworkers were shallow and transitory. Quite the contrary. Carver's engaging, intense personality attracted nearly as many people as his self-righteousness repelled. One such person was Harry O. Abbott, who served as Carver's traveling secretary in the 1930s, when the requests for Carver as a speaker reached their peak. Abbott not only arranged all of Carver's speeches and the accommodations for his trips but also traveled with him, becoming his close friend and confidante. After Abbott autographed Carver's copy of the 1931–1932 Tuskegee yearbook, appending his signature to several comments praising the professor, Carver responded with this letter.

May, 1-'33[22]

My esteemed friend, Mr. Abbott:-

We had talked and commented so much on the bound copies of the 1931–1932 Year-book that I neglected to look my copy through with more detail.

This evening I had occasion to consult it and turned to the autographed page for the first time.

I have read and re-read it several times and am not through yet. This page means more to me than all the rest of the book put together.

I have not added one half to your life that you have added to mine, you have encouraged and inspired me to do my best some times, in many, many ways.

Your ability to look through and size up man (in the generic) has always interested me very much.

I consider myself unusually fortionate in being able to get you to accompany me on these many humanitarian trips wc have made.

You not only releived me of every responsibility connected with the trip but was always on the alert and seemed especially happy when you could add anything to my personal comfort which I appreciate far more than I have words to express.

The excellent workmanship of the book speaks for itself and needs no further comments of mine.

Most sincerely yours,
G. W. Carver

Abbott left Tuskegee in 1937 to take a job in Chicago. Carver missed him immensely, and during the next six years, until his death in January 1943, he wrote more than eighty letters to Abbott. Often, particularly soon after Abbott's departure, only a few days elapsed between letters. Carver's prolific letter writing during these latter years is remarkable, given the fact that he was ill much of the time.

Carver visited Abbott in July 1937. Following the visit, and a letter from Abbott, Carver responded with a note that told of a longing for Abbott's companionship tempered only slightly by the national attention he was receiving from broadcast journalists.

July 31, 1937[23]

My beloved friend, Mr. Abbott:

Thank you very much for your splendid letter which reached me yesterday. It found me, however, pretty well down the line, as I have not been very well since my return from Chicago. The trip was very much too heavy for me. . . .

The chats that we had seemed to refresh me so very much, and if I could get them now I know that I would feel better. I am confident that some of my disabilities come from the fact that I can't see my friend, Harry. That may seem rather odd to you, nevertheless, it is true.

I did indeed miss you at the banquet, which was most unique and positively beautiful. And such a fine group was there. But through it all I wanted to see you there. . . .

Letters are pouring in, as you can well imagine, from all over the country with regard to the broadcast.

Leaving out the intensely personal part of it which affects me so much, I believe that there are opportunities for you there in Chicago, and I shall look forward to their development with more than passing interest. It maybe that I can get up there this fall. I am not sure yet.

The Metro-Golwyn-Mayer representative was here a few days ago. He flew down last Tuesday from New York, and stayed about an hour. I saw him for about five minutes. I have never seen a person more thoroughly impressed. He seemed to be almost overcome, and said that he would go back and work out something and submit it. . . .

Very sincerely yours,

G. W. Carver

There is a melancholic tone in the Carver-Abbott correspondence. The letters project Carver as a tired, sickly, often depressed old man who was becoming more and more aware of his own mortality. There are many references to mutual friends who had died recently, for example, and constant mentions of his own poor health. At times, Carver seemed to invite Abbott to do favors for him, almost as tests of the persistence of old-time loyalties—this from a man who continued to feel so unappreciated by the people who surrounded him. Carver told Abbott in September 1937:

September 15, 1937[24]

My beloved friend, Mr. Abbott:

This is just to extend to you greetings, as I have been thinking of you so much within the last few days.

I am just now beginning to be much more hopeful with reference to my physical condition, as I am getting better from a most aggravating something that began going on last March, and has annoyed me in increased fashion until last week. I have hit upon something that seems to be reaching the spot, and that terrible annoyance is leaving.

I trust that all is going well with you, and I am quite positive that you know that I miss you. . . .

Thank you so much for your constant interest in getting things that you think that I would like. This has always been characteristic of you, and has been appreciated much more than you will ever be able to know. I do hope that you can run across the product known as "Samp". . . . "Samp" is made in this way: The corn is hulled, dried, ground, and sifted. The fine particles that go through the first screen is called grits, as they are very fine. The second is coarser and called hominy grits. that is what we have here. I have not seen these in a very long time. Now the third sifting, or rather that which remains on the screen, is the "Samp", and is very coarse. This is the kind that I want. . . .

Very sincerely yours,
G. W. Carver

Carver relished the opportunity to tell Abbott about how much he was in demand, and the irony of being sought after elsewhere and unappreciated at home did not escape him.

January 28, 1938[25]

My beloved friend, Mr. Abbott:

Thank you so much for your fine letter. I am pleased to know that you are feeling quite well and that all is going well with you. I myself am not quite strong at present, as my heart is giving me a warning every little while to the effect that it may just get tired and stop for a long rest. . . .

Various and sundry are the comments that are going on yet with reference to the subjects you raise in your letter, and we must confess that they are most puzzling. I do think, however, that some lessons are being learned that will prove most valuable in the future. I myself am getting much satisfaction out of the fact that people right here on the grounds let a stranger come in and see in a few bits of observation what they had not seen in years with reference to myself. The little narrow prejudices blinded their vision. It always does. For the last few days we have been having genuine winter. I am now in Hooker's office sitting right against the radiator with my overcoat and cap on, and with one of these little curious electric heaters focused upon me.

It is perfectly terrible the way the people are pouring in, and the vast number of subjects that they want me to work out for them. And they think I can do every conceivable thing, and go everywhere.

I have just received a remarkable letter from Dr. Glen Clark of St. Paul, Minn., wanting me to come as his guest to St. Paul on April 5 or 6, all expenses paid, plus $100.00 honorarium, with definite provisions made by the president of the L. and N. R. R. for drawing room and every other comfort that can be provided.

He said that they would have absolutely no trouble in filling an auditorium which holds 2100 people. The occasion is the meeting of a very spiritual group that is arranging a series of lectures on bringing Christ into our lives during the Week before easter. I doubt whether I will be able to go, while I would like to see that section of the country.

I have much more to say, but I am told that I have to stop right now. I am pleased, however, that you and I know where you and I stand in so far as permanent friendship is concerned.

Yours very sincerely,
G. W. Carver, Director

One of the driving ambitions during the last years of Carver's life was the establishment of a museum in his honor at Tuskegee. Carver hoped that the museum would do more than simply keep alive his own memory; he wanted it to be a place that would inspire future scientists to "catch the spirit" and to continue the work he had begun. He had grandiose plans for this facility, but his hopes for it never materialized. In his judgment, Tuskegee officials were unwilling to spend the amount of money that a testimonial to himself warranted. The subject of the museum was a topic of discussion in Carver's letters to his old Tuskegee colleague, Harry Abbott. It must be remembered that Carver's health was poor during those last years and that lack of control in his right hand forced him to dictate his letters to a secretary. Hence, he was reluctant to speak too candidly against administration officials, knowing that his message might get back to them. Still, his disappointment came through, as is evidenced in this excerpt from a letter of November 1938 to Abbott.[26]

> The museum is taking on shape and looking rather pretty. It is not at all what Mr. Frye had planned first, as the first plans were very elegant and really just what I wanted, but they called for nearly twenty thousand dollars which was prohibitive at the present time and I felt rather than put it off any longer that we had better take what we could get. This would give us at least a start so that it could be added to from time to time if the persons into whose hands it falls will catch the vision and go ahead with it, but it must not be delayed longer for reasons that we do not need to discuss. I wish we could have carried out Mr. Frye's original plans. He is so confident and shows such a fine spirit.
>
> Things are moving along here I feel along the lines of the least resistence naturally.

Ten days later, Carver wrote again to Abbott, and although the message was still somewhat abstract, it was nonetheless more forceful and pointed.[27]

> I agree with you that some of the disparaging things round about here have died away, I hope for good. All of those that there is nothing to, and others that there is a little food for thought I hope will be corrected. . . . I was perfectly conscious of what you state with reference to the Museum, as it was hoped that it would be "killed in the borning", but Mr. Anderson is doing a fine job and I am very much pleased with it much to the disheartenment of a very few, which after all, doesn't make any difference.

The museum was, of course, finished before Carver's death, but the old scientist never forgot or forgave the poor treatment he thought he had received. Tuskegee Institute, which had in the beginning been a source of hope and joy for him, turned out to be a place of alternating frustration and anxiety, even to the end. Little wonder that he turned increasingly away from Washington's school for confirmation of his worthiness and validation of the reality of his association with the Divine.

FIVE

The Teacher as Motivator

Carver and His Students

This old notion of swallowing down other peoples' ideas and problems just as they have worked them out without putting our brain and origionality into it and making them applicable to our specific needs must go, and the sooner we let them go, the sooner we will be a free and indipendent people.

Geo. W. Carver 19 December 1898

WHILE CARVER HAD his problems with Tuskegee administrators and coworkers, his relationships with students were much more rewarding. He was a popular and effective teacher who possessed the ability to inspire and excite would-be scholars. Even while Booker T. Washington was growing increasingly frustrated with Carver's administrative bungling and bickering, he recognized his troublesome subordinate's rare gifts as a teacher. In a letter of chastisement written on 26 February 1911, Washington criticized Carver's shortcomings as a manager, but added: "I think I ought to say to you again that everyone here recognizes that your great fort[e] is in teaching and lecturing. There are few people anywhere who have greater ability to inspire and instruct as a teacher and as a lecturer than is true of yourself. . . ."[1]

Indeed, in 1910, University of Georgia chancellor Walter Barnard Hill offered this assessment of Carver as a teacher, after hearing him speak at a Tuskegee Negro Farmer's Conference: "That was the best lecture on agriculture to which it has ever been my privilege to listen." Chancellor Hill continued, explaining that Carver had "not only shown himself a master of the subject," but also was "possessed of pedagogical ability to impart it clearly and forcibly to others—a combination which is possessed by only five or six men in the entire country."[2]

Carver drew upon three unique talents as a teacher. First, he was genuinely interested in his students and made them feel that they were truly important to him. Second, he was excited about his subject matter, and he transmitted that excitement to his students. And third, his teaching methods combined the transmission of ideas with their practical application, a combination that many of his students found irresistible. Above all else, Carver wanted his students to learn to think creatively and independently.

Carver did not compartmentalize his life into "work" and "home" spheres. At Simpson College and Iowa State he became accustomed to spending virtually all of his time and energy with people and activities associated with the respective campuses. He expected to continue that pattern of behavior at Tuskegee. For the majority of his years there, he lived in Rockefeller Hall, a dormitory occupied by many of his students. A man with no family, he thought of Tuskegee students as his children, as is evidenced in this thank-you note he penned to a representative of the senior class after receiving a Christmas present. Note in this letter Carver's reference to himself as "your father." Students often called him "Dad" or "Daddy" in their letters to him.

January 9, 1922[3]

Mr. L. Robinson

I wish to express through you to each member of the Senior class my deep appreciation for the fountain pen you so kindly and thoughtfully gave me Christmas.

This gift, like all the others, is characterized by simplicity and thoughtfulness, which I hope each member will make the slogan of their lives.

As your father, it is needless for me to keep saying, I hope, except for emphasis, that each one of my children will rise to the full height of your possibilities, which means the possession of these eight cardinal virtues which constitutes a lady or a gentleman.

1st. Be clean both inside and outside.

2nd. Who neither looks up to the rich or down on the poor.

3rd. Who loses, if needs be, without squealing.

4th. Who wins without bragging.

5th. Who is always considerate of women, children and old people.

6th. Who is too brave to lie.

7th. Who is too generous to cheat.

8th. Who takes his share of the world and lets other people have theirs.

May God help you to carry out these eight cardinal virtues and peace and prosperity be yours through life.

Lovingly yours,

G. W. Carver

Carver's concern for his students extended beyond the classroom. He cared about the whole person and was known to give advice and encouragement on a host of nonacademic matters. Sometimes he even loaned students money. His students remembered his assistance with gratitude and affection. One student wrote to him in 1923: "I have thought of you many times since I left you. Pencil and paper can not express my thanks to you for your kindness and generous attention you gave me. I know I shall never be able to pay you and thank you for the interest you took in me."[4] In 1912 Carver interceded with Booker T. Washington for a student whose medical bills threatened his academic career:

December 13, 1912[5]

Mr. B. T. Washington:

   I am sending you a note and board-bill from Nonses Cohen, both of which are self-explanatory. You doubtless recall that this young man is the one who was crippled last summer, and has been unable to do any work or even to walk since, has been in the hospital for at least two weeks—in fact he is in there now. He is not improving. He has only $50.00 at his disposal, providing first the hospital bill is waived; if not, he has not even enough to pay this bill. He is the young man who you are trying to get into the hospital of which you spoke before you left. I wish so much that something could be done at once for him, as it is very embarrassing for him to be in this disabled condition and have these bills presented him without anything with which to pay them.

Yours very truly,

G. W. Carver

On the lighter side, Carver was a kidder and practical jokester who loved to romp and play with his "children." One of his favorite forms of play was to administer mock whippings and beatings to his students. Indeed, he continued this diversion long after students left the school. Many of his letters to former students mention whippings and beatings, such as this one to William F. Smith of the Pickens County, Alabama, Training School.

December 9, 1941[6]

My dear boy, Smithy:

   Your letter is indeed quite refreshing. I am so glad to hear from you, and I am happy to learn you are well located and that you like your work. I am very sure that the pupils all like you and will do just what you say. . . .

Now why you would say that you are not getting any whippings is beside the point when you know very well that your back is drawing interest. I have not heard from Bassett yet. That boy had better be getting a letter to me very soon or else his back won't hold shucks when I get him.

I hope you will have a wonderful holiday season and that you will help them out in their various plays and holiday activities. It will be a wonderful opportunity for you.

All the Tuskegee family send their highest regards and are so much interested in what you are doing.

With much love, I am

Very sincerely yours,
G. W. Carver

It is remarkable that even in his late life, Carver's letters to his former students evidenced a spirit of joking, as this note to Samuel Richardson, a longtime correspondent, makes clear.

November 2, 1940[7]

My dear Reverend Richardson:

It is almost with big elephant tears in my eyes, and I am always afraid that they will splatter around about me when I think of your condition. Although it is not entirely hopeless for sometimes you encourage me by showing a little symptom of intelligence as after you wrote those other two vile notes you did such a fine thing to hop on a train and get away just as fast as you could. I don't blame you for that, but nevertheless your condition is drawing interest now, and I think compound interest, with much emphasis placed on the *pound*.

Your fine wife who is your salvation and keeps you from going absolutely to wreck and ruin, along with the beatings that I can muster strength enough to give you from time to time as badly as it hurts me to whip, encourages me to try to save you. I think it will be a great feather in our caps. If at all possible, I believe that every letter that you write gets worse. . . .

I shall be most happy to see you when you come by, but if I could just get you not to talk such big talk and write such terrible letters, then you wouldn't have to hop on the train and leave. . . .

I think that you had better meander around and let me work on you some and see if I can help you. I know that Mrs. Richardson wil be very thankful to me for such help. Yet, after all of your shortcomings (and they are many), occasionally I can see a long one here and there, mostly there and not here.

Very sincerely yours,
G. W. Carver

The fact that Carver joked and frolicked with his students, however, did not mean that he took his responsibilities as a teacher lightly. Although Washington placed a great many duties on his brilliant but temperamental subordinate, Carver tried never to let those other duties detract from his teaching. In 1899 he told the principal, "In regard to the teaching I have always felt as you do about it, and therefore have stuck very close to it at the expense, frequently of the other, and more practical part. . . . For recreation I go out and hoe, pull weeds and set plants myself."[8]

In 1906, Washington asked Carver for suggestions about how to improve the Agricultural Department. Carver's response revealed that the welfare and training of students was of paramount importance to him. It also revealed his underlying unhappiness with things as they were at Tuskegee.

June 4-'06[9]

Mr. B. T. Washington,

In response to your request, I beg to submit the following, which to my mind, if carried out will greatly improve the conditions in the Agrl. Dept.

Harmony

The first and most essential is harmony between the instructors, while they are fussing the student is suffering.

The need of better organization.

(a) Two distinct branches of Agrl. work are in the mechanical Dept. If the Agrl. student comes in touch with them at all he must be transferred to another Dept.

The principals underlying these divisions are purely agricultural, which gives you a little school of Agr. in the mechanical Dept.

(b) All teaching should come under one person, gieving such a person time to visit all classes in a suggestive way, this has been attempted heretofore, but the person made responsible was given a full schedule of classes so that it was impossible to carry out the above suggestions effectively.

Get more students into the Dept.

(a) To the above I suggest that special dodgers be sent out at once offering inducements, and distribute these through the country districts as well as towns and cities. The Conference Agent and the operator of the Jesup wagon can be mighty forces in this particular.

(b) Sci. Agricultural year book on the order of those issued by Tennessee, New York, and a number of other states containing nothing but Agricultural reading, pictures, mottoes and the courses of study, should be printed separately from the catalogue and distributed so.

(c) More effort should be made to make conditions as comfortable and as attractive on the farm for the student as in any other department.

Now he has not so much as a few boards put up to protect him from a storm should one come up when he is in certain fields.

He is often up early in the morning and late at night for the Dept. and too frequently given no tangible consideration.

Get more people here every year who represent some successful endeavor in Agriculture to lecture to the entire student body, and especially urge the faculty to be present.

The second or third chapel service the Principal should give an Agrl. talk which will greatly assist the students in the choice of a trade.

More Graduates.

In our great desire to turn out a large number of people there are two dangers which must be guarded against Viz. the "Field hand" and the "Talking Agr." Neither is Tuskegees ideal, but one who not only works but knows why he works, and can tell why he works; this requires time and careful preparation.

To meet the requirements as teachers I do not see how you can afford to give one whit less. I try to get our seniors to take a post graduate course after completing the regular one.

There is however a system of short courses which colleges and universities give with a greater or less degree of success which Tuskegee might possibly consider with profit. Such as Dairying; Truck Gardening, Floriculture, Landscape Gardening; Poultry raising; Live stock; ect.

The applicant to come here and choose the division he wishes to make a specialty of and stay the few months or years, as the case may be, until he completes the work of that division and receive a certificate therefrom.

Respectfully submitted
Geo. W. Carver.

Carver realized that most of his students had been educated in inferior segregated schools and that when they left Tuskegee they would return to a society rampant with racism. As a consequence, he felt it was all the more important that they be trained adequately. Though he might joke with and tease his students, he was a harsh taskmaster who demanded a great deal of work. His teaching methods always combined laboratory and field work with lectures. Carver was a great believer in and proponent of the so-called "Nature Study Movement," which was popular throughout the country, especially in rural areas, during his early years at Tuskegee. This movement sought to get students out of the classroom and into the laboratory of the great outdoors, where they could encounter nature firsthand.[10] In 1902, he published, through Tuskegee Institute, a bulletin that provided direction on how to employ the Nature Study approach to education. The preface to that publication is reproduced below. Note Carver's no-nonsense approach to teaching in this publication.

Suggestions for Progressive and Correlative Nature Study[11]

To The Teacher

In the beginning, every teacher should realize that a very large proportion of every true student's work must lie outside the class room. Therefore, the success of these outlines will depend largely upon your efforts in their presentation.

The study of Nature is both entertaining and instructive, and it is the only true method that leads up to a clear understanding of the great natural principles which surround every branch of business in which we may engage. Aside from this, it encourages investigation and stimulates originality.

This common fault is apparent in all outlines pertaining to this work: Students invariably want to discuss the topic, rather than give you a direct answer. This is not permissible, neither what he nor she may think, unless their thoughts are based upon facts. There is nothing to be deplored more in the class-room than to hear a number of pupils pretending to recite, and constantly telling you what they think with reference to matters that the intellectual world has recognized as fact decades ago.

They Must Know It

See that each student is prepared with slips of plain white or manilla paper, for making sketches, and insist on their work being kept very nice and clean; securing such as may be worthy for exhibitive purposes.

The following method of grading has proven very satisfactory:
Grade (a)—The best.
"   (b)—Poor.
"   (c)—Rejected.
Every energetic student will aspire for Grade (a). This grading only applies to neatness as some will naturally draw better than others.

Correct English [usage is a] prominent feature of these exercises.

Experience has suggested the changes made herein, as a matter of convenience to both teacher and pupil.

The teacher is especially urged to see that every topic is taken up and finished as outlined, before permitting the student to pass to another, amplifying each topic or subject as your wisdom dictates: carefully staying within the educational scope of this leaflet. The introduction of technical terms should be avoided in all of these topics, as at this point the child's mind is not able to grasp and retain them. Gross characters that are easily seen, are all that is desired.

The following year, Carver wrote a lengthy letter to his colleagues in which he expressed concern about the quality of students being produced by Tuskegee and outlined the teaching techniques he believed would remedy the situation.

[Incomplete Date] 1903[12]

Fellow Teachers:

For some time it has been strikingly and often painfully apparent that our students did not reach our ideal in proficiency of work, thought, and character. Some of you have been kind enough and thoughtful enough to write me notes, and others to speak to me personally about it, all of which did not pass by unheeded, but like bread cast upon the waters, it has returned after many days thus.

(a) The mere matter of mechanical genius may be developed to any degree, but if the possessor leaves his saws, hammers, and the various tools belonging to the trade, scattered about here and there, it disqualifies him for the highest usefulness in so far as the above is true.

Just so with the farmer. If his tools are left scattered here and there over the field or wherever he stops using them, and it is found necessary to follow him up and keep reminding him of the flagrant neglect of the various duties in which he should take a special pride, it is quite evident that these faults will go with him when he leaves unless corrected here.

I suggest the following as a correction:

(1) That the theory and practice be placed upon the same uniform basis as far as marks are concerned. I have therefore allowed,—

100 to equal first rank.

75 to equal second rank.

50 downward to equal third rank.

The quarterly examinations would be given by the Industrial teachers at the end of the quarter, marked in figures and sent to the Director's office, where the final ranks would be made by putting the two or more marks together and dividing by the number of marks represented; eg;— if (A) gets 100 in theory under Mr. Bridgeforth or myself, and only 25 in practice under Mr. C. W. Greene, Mr. Owens, or Mr. Gordon, this would equal 125 divided by. 2, which would give him a mark of 62, which, according to the above system, would give him an ultimate rank of 2, and the failure would be on the practical side.

The marks in the practical work should cover the following points:

(a) The proper cleaning, housing, and general care of tools, implements, stock, harness, wagons, vehicles, and everything with which the student works. They should also include the amount of brain exhibited in his work,—ie, the more thoughtful and the greater amount of interest manifested in his work, the higher, of course, would be his mark. This mark would include also the man's character—as to ugly disposition, unreliability, falsifying and other forms of moral weakness.

To my mind, the above will save the teacher a deal of annoyance in the way of constantly following the student up to see that tools, apparatus etc. are properly put away, and the constant annoyance on account of unreliability etc. by placing the student upon his own responsibility and marking him in accordance with the efficiency and faithfulness to his trust.

It will also avoid the embarrassing situation of sending him out into the world with our signature and the defects only too apparent.

It will also correct the evil of certain students spending most or all of their time in other departments as is now being done. This will compel them to be one thing or the other.

We also think well of a suggestion made some months ago, advocating the coming together of the entire Agricultural faculty twice every month for the discussion of the students and the ways and means by which our work can be more effectually correlated.

Respy.

Geo. W. Carver

Carver obviously had a clear notion of how a Tuskegee student should behave. If he saw behavior he believed was unbecoming to a student, he could be counted on to correct it, regardless of the situation. Such was the case with an incident he reported to President Patterson in 1937.

September 21, 1937[13]

Dear President Patterson:

This is just to extend to you greetings, and to say that I have looked over the student body, have come in contact with a number of them, and have noted the way they have started off. I am very much pleased.

I can already see some of the effects of the remarkable charge that you gave to the teachers in faculty meeting last week. What you said could not have been improved upon, to my mind. It was clear, concise, sympathetic, and very much to the point. If all our teachers can grasp the meaning of it, and fall in line one hundred per cent, we will have before the year is out a perfect example of what is meant by "The Tuskegee Spirit"—the thing that you asked them about last year and gave them a hint as to what it was.

I was very much gratified this morning in the Cafeteria. I was taking my breakfast. A young man came in with his stocking cap on and walked to toe line. I called him over to my table and asked him to remove it. He thanked me there. And, much to my astonishment, came to the agricultural building and thanked me for calling his attention to it. It seems to me that this is "The Tuskegee Spirit", and when we get a body of students who catch that spirit with the aid of the teachers, we will have a school second to none.

I am most happy to see the way that we are starting off.

Yours very truly,
G. W. Carver

One of the clearest statements of Carver's methods came in another "Nature Study" booklet he wrote, this one in 1910. Although it was addressed primarily to teachers of younger students, Carver's method of combining theory and practice applied to students of all ages. The excerpts reproduced below reveal that he perceived the entire natural world as a classroom from which one can learn many lessons. For Carver, gardening allowed students to understand the relationships of all things in nature to each other. Planting a garden provided students with an opportunity to study more than botany; it introduced them to mathematics, history, climatology, economics, advertising, landscaping, conservation, entomology, geology, and a host of other disciplines. Carver left nothing to chance, either, as his minute detailing of instructions to teachers makes clear.

Nature Study and Gardening for Rural Schools[14]

(June, 1910)

The chief mission of this little booklet is that of emphasizing the following points:

1. The awakening of a greater interest in practical nature lessons in the public schools of our section.

The thoughtful educator realizes that a very large part of the child's education must be gotten outside of the four walls designated as class room. He also understands that the most effective and lasting education is the one that makes the pupil handle, discuss and familiarize himself with the real things about him, of which the majority are surprisingly ignorant.

2. To bring before our young people in an attractive way a few of the cardinal principles of agriculture, with which nature study is synonymous.

If properly taught the practical Nature study method cannot fail to both entertain and instruct.

It is the only true method that leads up to a clear understanding of the fundamental principles which surround every branch of business in which we may engage. It also stimulates thought, investigation, and encourages originality.

Who has not watched with delight the wee tots with their toy set of garden tools and faces all aglow with happiness and the yearning expectations of the coming harvest as they dug up the earth and dropped in a few seed or illy set an equal number of plants—with what joy and satisfaction they called it their garden, or with what enthusiasm they hailed the first warm days of spring with their refreshing showers which bespoke emphatically the opening of the mud pie and doughnut season, and how, even though they were water-soaked and mud-bespattered from top to toe, how very happy they were at the close of such a day's work. So on through the whole list of childish amusements. Instinctively, they prefer to deal with natural objects and real things. It is the abnormal child that will feel just as happy with a piece of mud from which to make its cooky or pie crust as a piece of real dough. Neither is there the same instructive interest in a lifeless, irresponsive bundle of cotton cloth, ribbon and what not in shape of a kitten, puppy, etc., as there would be in the real, live, beautiful little animal which responds to every caress and which mutually seems to share in their joys and sorrows, successes or failures. . . .

When to Begin

The age to begin teaching or interesting the child in the growing of plants for himself has puzzled many, but my observation and personal contact with the work proves that with a well-equipped teacher the wee tots, the kindergarteners are none too small. Figure 12 [a photograph of children working in a garden] represents a group in a flower garden. Their faces portray the happiness within the heart. I am sure the value of such instructive interest and development could not be questioned.

In connection with the above many are the lessons which may be taught in:—

Correlation

Nature study as it comes from the child's enthusiastic endeavor to make a success in the garden furnishes abundance of subject matter for use in the composition, spelling, reading, arithmetic, geography, and history classes. A real bug found eating on the child's cabbage plant in his own little garden will be taken up with a vengeance in the composition class. He would much prefer to spell the real, living radish in the garden than the lifeless radish in the book. Likewise he would prefer to figure on the profit of the onions sold from his garden than those sold by some John Jones of Philadelphia.

Partnership

It has been the experience of many teachers that it works well to have two, three or four children form a partnership, under written contract, who will be assigned by the teacher to one of the little plots set apart as an individual garden. . . .

The contract will prove valuable only so far as the teacher makes the children understand just what a contract means, its binding effect in the business operations of the garden and the suffering or loss, regardless of excuses, to the person or persons who fail to come up to the stipulations.

The co-operation plan can also be taught by having one child responsible for a garden plot, but as concerns walks, borders, etc., equally responsible for the corresponding straightness and cleanliness throughout.

How to Begin Gardening

In response to many inquiries as to the best way to begin, we beg to say that there are many—all possessing some merit, but the following has proven most satisfactory with us:

1. As an introduction, a fifteen-minute lecture or general discussion on garden work should be given, defining the [kinds of gardens that can be grown]. . . .

Selection of the Site and Why

The out-door garden should be a plot of land near the school building. The garden should be used as any other class room. Hence, the closer it can be brought to the other places of recitation the better.

Laying off the Individual Plots

The plots or individual gardens should represent an easy fractional part of an acre, e.g., one-twentieth, one-fortieth, one-fiftieth, etc. A walk two feet wide should surround the entire plot. The individual gardens should be separated by paths not exceeding two feet.

Selection and Care of Tools

A hoe, rake and spade are essential. A hand weeder (which can be easily made by bending a piece of hoop-iron triangular or any other shape desired, fitting a handle to it and sharpening the edge) could be used. A line to insure straight rows and dibber for making holes to set plants will be found inexpensive, convenient and useful. A set of tools for each child is ideal and desirable, but fewer can be made to answer by arranging the work so that some will be using a hoe while others are raking, spading, laying off their plots, etc. Tools should be carefully cleaned and dried each time they are used and put in the place set apart for them. This should be an essential part of the exercises in the garden.

Preparation of Soil

After the land has been cleared of objectionable things such as stumps, stones, etc., it is ready to be spaded or plowed up deeply and thoroughly. Turn every furrow or spadeful of earth upside down, following this process with a thorough chopping over with a hoe or harrow until all of the large clods are broken. Finish with the rake. . . .

Selection and Testing of Seeds

Some time should be given to the study of garden seeds. The child should be taught how to select large, plump and well developed seeds and plant them in a dish of fresh sand, moistened with clean water and kept in a warm place as a test for vitality, or germinating powers. A box of moistened earth kept in a warm place may also serve for the same purpose. A number of interesting and valuable mathematical exercises can be developed by planting a definite number of seeds and calculating the percentage of loss or failures. That is, if sixteen seeds are planted and eight germinate the per cent. is one-half or fifty. If only four sprout the loss or failure is three-fourths or seventy-five per cent., etc. Of the first lot of seed one would have to plant twice the normal amount if a stand was expected; of the second three times the normal amount should be planted. . . .

Planting

Many interesting and valuable lessons can be brought out relating to the size of seed, ease of germination and the depth to plant. The last depending largely upon the character of soils and the weather conditions. It should be kept in mind that seeds should be planted more shallow in heavy clay loam than sandy; and deeper in dry weather than in wet. This especially applies to seeds that germinate quickly.

This is an old rule but a very good general one: Plant all seeds to a depth of three times their greatest diameter. A number of seeds of different kinds should be measured by the pupils until the principle is thoroughly understood.

Cultivation

The following simple reasons for this operation should be enlarged and dwelt upon until it is made clear to every pupil.

(a) How cultivation destroys weeds and why destroy them.

(b) Lets water into the soil and prevents washing; how, and why this is desirable.

(c) Permits the roots to go deep into the earth and to reach out long distances in every direction. (NOTE—For what are they seeking?) (Teacher explain.)

(d) Lets air and sunlight into the soil. (NOTE—For what reason?) (Teacher explain.)

Reaping the result of any well directed effort is more or less interesting and many are the good wholesome lessons which can be taught as to

(a) When things are ready.

(b) The manner and quality of the product.

(c) The proper weather conditions.

(d) Saving the crop under adverse conditions, etc.

Marketing

While the harvest is full of interest from beginning to end, nothing inspires and encourages like the beginning or swelling of a bank account or the prompt payment of any debt which becomes due as a consequence of our business operations. With what enthusiasm the child watches the growth of the credit side of its garden business. . . .

Money Value of Different Garden Crops

As the child markets his produce he will have the opportunity to compare the money value of his garden crops. For instance, he may find by comparison that the quicker his crop is grown and off the greater will be the amount made on his small area of land. Comparisons may be made with slower maturing crops that are grown in the fields.

In this connection some valuable lessons may be taught as to how vegetables, grains, fruits, etc. (for which there is no paying market) can be made to pay by feeding them to chickens, hogs, cattle, etc., etc., or, in other words, turning them into pork, milk, butter, beef, eggs, etc. for which there is always a market.

Climate

During the year many important climatic changes can be noted as follows:

(a) The washing of the soil by heavy rains and how it impacts certain kinds and how it effects the growth of plants.

(b) The effect of excessive heat and cold upon plant growth.

(c) How some of these conditions can be greatly modified and overcome by tillage, and other soil manipulations.

A few Insects of the Year

During the entire year insects can be profitably studied. . . .

How to Classify Insects

Practically insects are divided into two great classes:

(a) Those that eat the leaves and other parts of the plant upon which they feed.

(b) Those that simply suck the juices from the plants upon which they feed.

The following remedies (which are poisons and must be used with reasonable care as to keeping it away from the children, etc.,) will exterminate or hold in check those insects which are most troublesome in our gardens. . . .

Remedies

Paris Green, this is the universal remedy for all kinds of biting or chewing insects. It comes in a fine dusty green powder and may be used in that way or mixed with water. . . .

All insects that suck must be killed by contact remedies, i.e., remedies when applied to the insect's body shuts up its breathing pores and smothers it to death. The favorite remedy here is kerosene emulsion. . . .

Stored Grain

The pea, bean and grain weevil of all kinds may be destroyed by bisulphide of carbon, a vile smelling liquid, which is very volatile (going into gas readily), it is very explosive, and must be kept away from fire. . . .

Trap Crops

All insects love young tender, juicy plants, and will feed upon them in preference to old ones, many planters take advantage of this fact and plant patches alongside or rows in between the old crop. Most of the insects will attack the young crop. They can here be easily destroyed by spraying with pure kerosene or boiling hot water.

Turnips, mustard, rape, radishes, etc., are the trap crops most commonly used.

Plowing

On the destruction of insects the time and manner play no small part.

As to the time, plow just as early in the fall as the crop can be gotten off, the earlier the better, many white grubs, wire worms harlequin cabbage beetles, corn worms, army worms, pupae, grub worms, and other insects will be destroyed.

As to the manner, deep (9-inch) plowing is preferable at all times, the soil should be turned completely upside down. . . .

Making Fertilizers

Too much stress cannot be laid upon this important item. All the weeds, grass, leaves, pine tag, wood ashes, old plaster, lime, old clothing, shoes, broken up bones, feathers, hair, horns and hoofs of animals, swamp muck, etc., should go into the compost heap.

Select a convenient place for the heap. Hollow out the ground into a sort of pit or basin to prevent the heap from leaching and therefore wasting more or less of its valuable fertilizing constituents. Put down a layer of leaves, etc., say from 6 to 8 inches deep. On top of this a layer of swamp-muck to the same depth. Let this alternating process continue until your heap is as high as you desire. After covering with a rough shed or shaping the top somewhat like a potato hill to turn the bulk of water, let all remain until thoroughly rotted. Barn-yard manure, cotton seed, etc., may be mixed in. . . .

To obtain the best results, the barnyard manure should be well rotted, and the commercial fertilizer thoroughly mixed with it before applying to the land. . . .

How to Dig and Set Trees

The first essential in getting a tree to live is to dig it properly, which means the getting of just as many of the roots as possible, the more earth you can save clinging to the roots in lifting, the better.

A hole from 16 to 18 inches larger than the ball of earth or the longest mass of roots on the tree should be dug out from 2 to 3 feet deep, fill with well rooted barnyard manure and leaf mould, to the desired depth, chop it into the other earth well with a spade, set the tree and firm the earth well around it, water thoroughly and fill with earth nearly to the top. The same method recommended for the digging and setting of trees applies equally well to shrubs and vines. . . .

How to make a Lawn

Since the Bermuda grass is the best one in this section for lawn making, I shall confine my suggestion almost, if not wholly, to it.

The site for the lawn should first be gotten in order by removing stumps, stones, filling washes and making the contour you wish, then plow, improve and prepare exactly the same as for the garden.

If a quick effect is desired lay the sods close together, if not run light furrows from 1 to 2 feet apart, drop in pieces of sod and cover lightly with earth. Excellent lawns can be made from early spring to midsummer in this way.

Flower Beds

In addition to splendid trees, and lovely grass new well arranged flower beds will add much to the beauty of the surroundings. These in number, size, shape, etc., may be just what fancy dictates. The ground should be made very rich, in fact, just as recommended for the garden. . . .

Window and Veranda Boxes

There is nothing that adds so much of real beauty, cheerfulness, and instructive inspiration, for the small outlay as a few pots or boxes of well grown plants.

Even though the schoolhouse be of logs, "clapboards" or what not, it can be transformed into a bright and cheery spot, with a little whitewash inside and out and a nice pot of wandering Jew, Kennelworth ivy, sweet potato vines, morning glories, etc., as hanging baskets in the window, on the veranda or especially prepared shelves, which the children will delight to make. . . .

Beautifying the School Grounds

No place can be called a school in the highest sense that has no pictures on the wall, no paint or whitewash on the buildings, either inside or outside, no trees, shrubs, vines, grasses or properly laid out walks and paths, which appeal to the child's aesthetic nature, and sets before him the most important of all secular lessons—order and system. With this end in view a few suggestions are here offered with the hope that every teacher will put forth strenuous efforts to see that his school grounds are made just as attractive as possible with the native trees, shrubs, vines, etc., about you. . . .

No doubt, Carver's methods and sincerity touched the lives of many students in his decades of teaching at Tuskegee. Although he spent less time in the classroom after Washington's death in 1915, he still taught, sometimes in a regular classroom, but more often through Farmers' Institutes, the Short Course in Agriculture, and summer schools. His teaching had a tremendous impact, as scores of his former students left Tuskegee to spread the word of what they had learned. Back in Carver's home state of Missouri, for example, a former Carver student, Nathaniel C. Bruce, established a school that he dubbed "the Tuskegee of the Midwest." Bruce remained in contact with Carver and tried to implement the latter's ideas in a school that lasted for more than half a century.[15]

It is impossible to read Carver's correspondence and his other writings without being impressed by his devotion to students and his love of teaching. Undoubtedly, he had a major effect on those students with whom he came into contact. But, somehow, the rewards were not enough. Bitter feuding with colleagues, his increased concentration on research, and an ever-increasing demand for him as a public speaker drew Carver away from the students he had gone to Tuskegee to help. Ultimately, Carver turned elsewhere in his unending search for the recognition he felt he so richly deserved and was so unfairly denied.

SIX

The Scientist as Servant

"Helping the Man Farthest Down"

The primary idea in all of my work was to help the farmer and fill the poor man's empty dinner pail.... My idea is to help the "man farthest down", This is why I have made every process just as simple as I could to put it within his reach.

Geo. W. Carver 16 January 1929

GEORGE WASHINGTON CARVER's early years of controversy at Tuskegee set the tone for a troubled relationship with his coworkers that lasted for the remainder of his career. Finding that he was unappreciated and unrewarded on campus, he turned to the outside world for praise and adulation.

But he also turned to the outside world to be of service to the people he felt that God had called him to serve. The suffering of poor southern farmers, black and white, in what he called "the lowlands of sorrow," greatly troubled him. Scientific agriculture, he was convinced, could be their salvation. Coincidentally, it might also be the vehicle by which his worth to the world would become widely recognized.

Carver envisioned the scientist as a person who unlocked the mysteries of the universe in order to improve the quality of life for everyone, particularly the poor and underprivileged. He believed that nothing existed without purpose. The job of the scientist was to discover the purpose and publicize its possible benefits for mankind.

Carver quickly understood that the farmers he sought to help were being victimized by their region's reliance on cotton, by the soil erosion caused by the heavy planting of cotton, and by the absence of the use of fertilizers to rejuvenate the soil. He set out to try to teach farmers how to revitalize the soil and how to raise diverse crops, including new foodstuffs that could supplement the limited diet of poor southerners.

In 1899, he outlined the problem, and a proposed plan of action to remedy it, in an essay titled "A Few Hints to Southern Farmers." The essay was published in *Southern Workman* in September 1899.[1]

The virgin fertility of our southern soils and the vast amount of cheap and unskilled labor that has been put upon them, have been a curse rather than a blessing to agriculture; this exhaustive system of cultivation, the destruction of forests, the rapid and almost constant decomposition of organic matter, and the great number of noxious insects and fungi that appear every year, make our agricultural problem one requiring more brains than that of the North, East, or West. Other occupations are holding out inducements to young men and women ready to grapple with life's responsibilities and the average southern farm has little more to offer than about thirty-seven percent of a cotton crop selling at four and a half cents a pound and costing five and six to produce, together with the proverbial mule, primitive implements, and frequently a vast territory of barren and furrowed hillsides and wasted valleys. With this prospect staring them in the face, is it any wonder that the youth of our land seek some occupation other than that of farming?

Yet we have a perfect foundation for an ideal country: we have natural advantages of which we may feel justly proud. Our soils are by nature rich and responsive, and our climate is so varied that the most fastidious may be suited. With a little effort we can have green forage for our stock the entire year; and we can raise with ease all the vegetables of the temperate, and many of those of the tropical regions. In truth the colder sections of the country should depend even more largely than they do upon our farms for their supply of early vegetables and fruits. This ideal will be realized in proportion to the rapidity with which we convert our unskilled and non-productive labor into that which is skilled and productive.

I see no logical reason why, with proper education and proper economy, we cannot in time make more butter and raise more valuable stock. In our by-products of the cotton seed, and in our wealth of leguminous plants we have the cheapest and best food in the world for the production of the best quality of butter and cheese; as to beef, pork and mutton we could enter into the keenest and sharpest competition with any other states.

There are certain things that I should like to suggest. We should greatly increase, both in quantity and in quality, all of our farm animals; we should sacrifice the razor-back hog the long-horned steer, and the scrub cow, which last animal requires the same food and care as a thorough-bred cow, and in return gives two or three quarts of two percent milk per day. We should also sacrifice a goodly number of the worthless puppies that are in evidence in too many dooryards, and put two or three sheep in their places.

It is pleasant to note that in many sections of our country agriculture in the primary grades is part of the compulsory curriculum. Nature study leaflets are being issued, and carefully planned courses of study are rapidly gaining place in many of our best colleges and academies, familiarizing the people with the commonest things about them, of which fully two-thirds are surprisingly ignorant. They know nothing about the mutual relationship of the animal, mineral and vegetable kingdoms, and how utterly impossible it is for one to exist in a highly organized state without the other. Our young people must be encouraged to take advantage of these opportunities. Send your boy to the best agricultural school within your reach, and let him take as much of the two, four, or six years' course as he can. Likewise, let your daughter go and learn the techniques of poultry raising, dairying, fruit growing, and landscape gardening. Be a frequent visitor at your nearest experiment stations; ask many questions, and note carefully every successful experiments then go home and try to put the same to practice on your own farm. Read their bulletins, and take one or two good agricultural papers; and a stream of prosperity will flow in upon you by reason of the application of this knowledge, that will be to you a very agreeable surprise. In conclusion I would say, attend with your family every farmers' meeting or conference, for they are powerful education factors, and are the keys to the golden doors of your success.

In 1909, Carver again challenged the South's over-reliance on cotton production, and penned an essay in the February 13, 1909, edition of *The Colored Alabamian* that he hoped would drive home the point. The essay was titled simply "Prof. Carver's Advice to Farmers," with a subtitle, "Cheap Cotton."[2]

Another years has come and gone. What has it meant to us? Will the mistakes of 1908 be repeated in 1909? We trust not. A few days ago I took dinner in a country home, this was the bill of fare, flour from Minnesota, coffee from Brazil, macaroni from Italy, cheese from Wisconsin, bacon from Kansas, and cake made from eggs purchased at the store.

In passing through the country I find a number of so called farmers without a garden of any description, not even a collard nor an onion.

I also saw farmer sell a bale of cotton for 7 cents per pound and buy side meat for 15 ½ cents per pound. The next week I visited another home.

They had cow peas, sweet potatoes, fruit, chicken, pork and greens, milk butter and cream, eggs, lettuce, radishes and onions, peaches, pears, corn bread and wheat bread. The corn was raised at home and ground at a nearby water mill and the receipts from the sale of butter and eggs purchased the flour. In fact everything on this table was purchased at home even to the bunch of flowers that adorned the center of the table.

It is needless to comment on these two farmers, the first was literally living out of the store, the later was not only living at home but had a surplus to sell.

In looking into the former's conditions more closely, I found that he had planted everything in cotton, even to the front yard around his home. So that he had nothing to eat except what he bought and nothing to sell except cotton.

The other farmer raised plenty of corn, peas, potatoes, garden vegetables, fruit, fowl, etc. and has a few bales of cotton he is holding for better prices.

Let me not make the mistake of putting everything in one crop, any one crop system is ruinous to the small farmer.

Remember that a farmer to be happy and prosperous, must raise not only what he eats at home but must have a surplus to sell.

Also to be prosperous he must have a little money coming in all of the year. Now is the time to make your plans. Sit down and talk the matter over with your wife, and decide that so many acres shall be put in corn, potatoes, peas, and a certain amount set apart shall be planted in such and such things. Also that we will keep at least one cow, some hogs, chickens etc. Put this plan on paper, work to it and you will just begin to find out that there is real joy and satisfaction in being a farmer.

From the beginning of Carver's tenure at Tuskegee, he was concerned with restoring the fertility of the nutrient-depleted soils of the South. Although he realized that commercial fertilizers could provide a short-term solution to the problem of worn-out soil, he knew that most of the farmers with whom he worked could not afford to purchase commercial fertilizers. In a 1911 letter to Booker T. Washington, he laid out an alternative solution.

January 26, 1911[3]

Mr. B. T. Washington

Again replying to your note regarding waste I beg to refer to my reply of January 5th to your same request. In closing my report I say I have in mind a matter that I think is almost or quite equal to the loss of coal, etc. I had in mind the enormous loss of manure in its broadest sense; e.g., beginning just below the hospital and following the little ravine down so far as the Harry Johnson tract, there are hundreds of tons of the finest kind of manure, which consists of decayed leaves, dead animals, decayed night soil, animal manures that have washed from the hillsides, etc., etc. These deposits have been accumulating for years. At this late date it is hardly necessary to discuss the value of such manure. It is recognized as the acme of manures and that which will build up a soil permanently. It is a source of the keenest regret that we are getting away from this important fact.

Let me quote from a circular on farm fertilizers recently issued by Dr. Knapp. He is thoroughly sound in principle and every farmer should read it and strive to carry out its principles: "Commercial fertilizers are costly; their excessive use tends to hasten the depletion of the soil, and they should never be considered a substitute for green crops or barnyard manure".

It is true that some of this manure does not contain as much fertilizing ingredients as if it had been composted, yet it will pay to save it. We should look to the permanent building of our soils. We know that commercial fertilizer will stimulate and for a while produce good results in the way of vegetation but by and by a collapse will come, as the soil will be reduced to practically clay and sand. The crying need of nearly every foot of land we have in cultivation is vegetable matter (humus) and every possible means at our command should be excercised to supply this need.

The above does not mean that we are to use no commercial fertilizer, but supplement our home manures with them, which will not only give us as good results, but permanently improve the soil.

It was my intention to analyze a number of samples of this manure and show you just how much plant food it contained, and how much commercial fertilizer would be required to bring it up to the standard, but I am not yet in a position to do this, and I do not think such an important element of waste should go on another day without steps being taken to correct it.

RECOMMENDATIONS

1st That where manure has accumulated on the hillsides, bottoms, etc., to any appreciable degree that it be raked up and hauled out on the farm and various places where needed.

2nd That the pits where the night soil, etc., has thoroughly decayed be treated in the same way.

3rd That the stock be fed or penned at night where the manure can be saved.

4th That composting, or its equivalent, be put into effect at once, whereby the maximum amount of the manure's fertility can be saved.

With the increased number of animals we are getting and the non-expansion of our cultivated lands, the expenditure for commercial fertilizers should grow less every year. I am planning to carry out a number of experiments on the Experiment Station to prove its value.

Respectfully submitted

Geo. W. Carver,

Director Dept. of Research & Experiment Station

George Washington Carver hoped to improve the lives of poor southern farmers by helping them to become more self-sufficient, by teaching them to restore the vitality of the soil, by introducing new crops that would allow for alternatives to cotton production, and by enriching the southern diet by adding new items to farmers' food supply, products that they could grow themselves.

Toward the accomplishment of these goals, Carver began experimenting with sweet potatoes in 1897. The following year, he published a bulletin titled "Experiments With Sweet Potatoes." During World War I, in the face of war-related food shortages, the U. S. Department of Agriculture requested Carver's help in demonstrating how to produce sweet potato flour, to offset shortages of wheat flour, and to produce sweet potato breakfast foods.

Similarly, Carver began experimenting with peanuts in 1907. He was interested in the peanut as a soil rejuvenator, but also as a supplement to the protein-lacking diet of the African American southern farm family. His publication "How to Grow the Peanut and 105 Ways of Preparing It for Human Consumption" appeared as Experiment Station Bulletin No. 31 in 1916.

Carver's "discoveries" of the multiple uses of the peanut were what brought him lasting fame. His testimony before the House Ways and Means Committee in 1921, as an advocate for a protective tariff in favor of the lowly peanut, thrust him into the spotlight as a spokesman for scientific agriculture. There is perhaps no better illustration of his vision of the practical role to be played by the scientist in American society than that testimony. Accordingly, it is reproduced below in its entirety.[4]

The CHAIRMAN [Joseph W. Fordney, R-Michigan]. All right, Mr. Carver. We will give you 10 minutes.

Mr. CARVER. Mr. Chairman, I have been asked by the United Peanut Growers' Association to tell you something about the possibility of the peanut and its possible extension. I come from Tuskegee, Ala. I am engaged in agricultural research work, and I have given some attention to the peanut, but not as much as I expect to give. I have given a great deal of time to the sweet potato and allied southern crops. I am especially interested in southern crops and their possibilities, and the peanut comes in, I think, for one of the most remarkable crops that we are all acquainted with. It will tell us a number of things that we do not already know, and you will also observe that it has possibilities that we are just beginning to find out.

If I may have a little space here to put these things down, I should like to exhibit them to you. I am going to just touch a few high places here and there because in 10 minutes you will tell me to stop.

This is the crushed cake, which has a great many possibilities. I simply call attention to that. The crushed cake may be used in all sorts of combinations—for flours and meals and breakfast foods and a great many things that I have not time to touch upon just now.

Then we have the hulls, which are ground and made into a meal for burnishing tin plate. It has a very important value in that direction, and more of it is going to be used as the tin-plate manufacturers understand its value.

The CHAIRMAN. If you have anything to drink, don't put it under the table.

Mr. CARVER. I am not ready to use them just now. They will come later if my 10 minutes are extended. [Laughter.]

Now there is a rather interesting confection.

Mr. [John N.] GARNER [D-Texas]. Let us have order. This man knows a great deal about this business.

The CHAIRMAN. Yes, let us have order in the room.

Mr. CARVER. This is another confection. It is peanuts covered with chocolate. As I passed through Greensboro, S.C., I noticed in one of the stores that this was displayed on the market, and, as it is understood better, more of it is going to be made up into this form.

Here is a breakfast food. I am very sorry that you can not taste this, so I will taste it for you. [Laughter.]

Now this is a combination and, by the way, one of the finest breakfast foods that you or anyone else has ever seen. It is a combination of the sweet potato and the peanut, and if you will pardon a little digression here I will state that the peanut and the sweet potato are twin brothers and can not and should not be separated. They are two of the greatest products that God has ever given us. They can be made into a perfectly balanced ration. If all of the other foodstuffs were destroyed—that is, vegetable foodstuffs were destroyed—a perfectly balanced ration with all of the nutriment in it could be made with the sweet potato and the peanut. From the sweet potato we get starches and carbohydrates, and from the peanut we get all the muscle-building properties.

Mr. [John Q.] TILSON [R-Connecticut]. Do you want a watermelon to go along with that?

Mr. CARVER. Well, of course, you do not have to have it. Of course, if you want a desert, that comes in very well, but you know we can get along pretty well without dessert. The recent war has taught us that.

Here is the original salted peanut, for which there is an increasing demand, and here is a very fine peanut bar. The peanut bar is coming into prominence in a way that very few of us recognize, and the manufacturers of this peanut bar have learned that it is a very difficult matter to get a binder for it, something to stick it together. That is found in the sweet potato sirup. The sweet potato sirup makes one of the best binders of anything yet found. So in comes the sweet potato again.

Then we have the peanut stock food. This is No. 1, which consists of the ground hay, ground into meal, much the same as our alfalfa hay, which has much of the same composition as our alfalfa hay, and we are going to use more of it just as soon as we find out its value. So that nothing about the peanut need to be thrown away.

Here is peanut meal No. 2. That can be used for making flours and confections and candies, and doughnuts, and Zu-Zus and ginger bread and all sorts of things of that kind.

Here is another kind of breakfast food. This is almost the equal of breakfast food No. 1. It will also have a considerable value in the market.

Here is a sample of peanut hearts. Now, it is not necessary for me to say that the peanut hearts must be removed from the peanut before many of the very fine articles of manufactured products can be made, such as peanut butter and various confections that go into candies and so forth.

Now these peanut hearts are used for feeding pigeons. Pigeon growers claim that it is one of the very best foods that they have found.

Now there is an entirely new thing in the way of combinations. It is a new thing for making ice cream. It is a powder made largely from peanuts, with a little sweet potato injected into it to give it the necessary consistency. But it is far ahead of any flavoring yet found for ice cream. It is a very new product that is going to have considerable value.

You know the country now is alive looking for new things that can be put out in the dietary. Here is a meal, No. 1. That is used for very fine cooking and confections of various kinds. I will not attempt to tell you how it is made.

Here is another thing that is quite interesting. This consists of the little skins that come off of the peanut. These skins are used in a very great many ways. They are used for dyes. About 30 different dyes can be made from the skins, ranging from black to orange yellow. And in addition to that they contain a substance very similar to quinine, and they are using a good deal of this now as a substitute for quinine, and physicians find it is quite attractive. Just how far it is going to affect the medical profession it is difficult to say, but I am very much interested in it myself, because the more I work with it the more I find that it is interesting and has great possibilities.

Here is another type of breakfast food quite as attractive as the other two. It is ready to serve. All that is necessary is to use cream and sugar, and very little sugar, because it is quite sweet enough.

Here is breakfast food No. 5. That contains more protein than any of the others. One of them is a diabetic food. If any of you are suffering from that disease, you will find one of these breakfast foods very valuable, because it contains such a small amount of starch and sugar.

Here is a stock food that is quite as attractive as any now on the market. It consists of a combination of peanut meal and peanut hay, together with molasses, making a sweet food of it, and chinaberries. The chinaberry has a great many medicinal properties, such as saponin and mangrove, and many other of those peculiar complex bodies that make it an especially valuable product that we are going to use as soon as we find out its value. All kinds of stock eat them with relish, and thrive upon them, and when they are added to these other foodstuffs, it makes a tonic stock-food. I have tried that out to a considerable extent on the school grounds, and I find that it is a very fine thing indeed.

Here is another breakfast food that has its value. I will not attempt to tell you, because there are several of these breakfast foods that I will not take the time to describe, because I suppose my 10 minutes' time is about up. Of course I had to lose some time in getting these samples out.
The CHAIRMAN. We will give you more time, Mr. Carver.
Mr. CARVER. Thank you.
Mr. GARNER: Yes. I think this is very interesting. I think his time should be extended.

Mr. CARVER. This is really the chinaberry. The other was not the chinaberry. It consists of a composition of ground peanuts and the peanut hay and the peanut bran, and so forth, made up into a balanced ration.

Now a great many people do not know the chinaberry. I have been very much interested in it. In fact I dug up a bulletin here a few days ago and found out that an English gentleman had gone over into Morocco and was attracted by the great yield of chinaberries and had made rather an exhaustive analysis of them and was advocating their use. However, we in Alabama advocated the use of the chinaberry several years ago.

Mr. [Henry T.] RAINEY [R-Illinois]. Do we produce in this country the chinaberry?

Mr. CARVER. Oh, yes.

Mr. GARNER. Yes, a great deal of it.

Mr. CARVER. And it is one crop that is infallible year after year. We understand that the chinaberry belongs to the same group of vegetation that the mahogany tree belongs to, and all of them have great medicinal properties.

Mr. RAINEY. Do we import any chinaberries?

Mr. CARVER. No, sir; we do not import any chinaberries, because we grow them in abundance here. Of course they are not used now.

Mr. GARNER. They are not used to any great extent.

Mr. CARVER. No, sir; they are not used to any great extent, but I am just as confident as that I stand here that they will be used as soon as we find out their value.

Mr. RAINEY. You don't need any tariff on them, do you?

Mr. CARVER. No, sir; we don't need any tariff on them down there. We need to know their value and to profit by their value.

Mr. RAINEY. The use of peanuts is increasing rapidly, is it not?

Mr. CARVER. I beg your pardon?

Mr. RAINEY. The varied use of the peanut is increasing rapidly?

Mr. CARVER. Yes, sir.

Mr. RAINEY. It is an exceedingly valuable product, is it not?

Mr. CARVER. We are just beginning to learn the value of the peanut.

Mr. RAINEY. Is it not going to be such a valuable product that the more we have of them here the better we are off?

Mr. CARVER. Well, that depends. It depends upon the problems that these gentlemen have brought before you.

Mr. RAINEY. Could we get too much of them, they being so valuable for stock foods and everything else?

Mr. CARVER. Well, of course, we would have to have protection for them. [Laughter.] That is, we could not allow other countries to come in and take our rights away from us.

I wish to say here in all sincerity that America produces better peanuts than any other part of the world, as far as I have been able to test them out.

Mr. RAINEY. Then we need not fear these inferior peanuts from abroad at all? They would not compete with our better peanuts?

Mr. CARVER. Well, you know that is just about like everything else. You know that some people like oleomargarine just as well as butter, and some people like lard just as well as butter. So sometimes you have to protect a good thing.

Mr. RAINEY. We have not any tariff on oleomargarine.

Mr. CARVER. I just used that as an illustration.

Mr. RAINEY. But to still carry out your illustration further, oleomargarine is in competition with butter.

Mr. CARVER. I believe that the dairy people want it there.

Mr. RAINEY. They never asked for a tariff on oleomargarine.

Mr. [William A.] OLDFIELD [D-Arkansas]. But they did put a tax on butter.

Mr. GARNER. And they did use the taxing power to put it out of business.

Mr. CARVER. Oh, yes. Yes, sir. That is all the tariff means—to put the other fellow out of business. [Laughter.]

The CHAIRMAN. Go ahead, brother. Your time is unlimited.

Mr. CARVER. Now, I want very hastily to bring before you another phase of the peanut industry which I think is well worth considering. Here a short time ago, or some months ago, we found how to extract milk from peanuts. Here is a bottle of milk that is extracted from peanuts. Now, it is absolutely impossible to tell that milk from cow's milk in looks and general appearance. This is normal milk. The cream rises on it the same as on cow's milk, and in fact it has much of the same composition as cow's milk.

Here is a bottle of full cream. That cream is very rich in fats, and can be used the same as the cream from cow's milk.

Mr. OLDFIELD. You made a mistake there, didn't you. you haven't got the milk mixed up, have you?

Mr. CARVER. No, sir. I may have made a little mistake there, but nevertheless the result is the same. That is what I say. No. 1 might be No. 2 because I am running over them very hastily.

Mr. OLDFIELD. Don't you think we ought to put a tax on that peanut milk so as to keep it from competing with the dairy product?

Mr. CARVER. No, sir. It is not going to specially affect the dairy products. We don't mean that it shall affect dairy products, because it has a distinct value of its own, and can be put right alongside of the dairy products, and if it is—

Mr. OLDFIELD. Did you say it was used for the same purpose as dairy products?

Mr. CARVER. It is used for the same purposes, yes, sir.

Mr. OLDFIELD. What is the reason it won't displace dairy products, then? Every time you use a pint of it will that not displace the dairy products?

Mr. CARVER. We do not now make as much milk and butter as we need in the United States, and then there are some people that would choose dairy products rather than these, and vice versa. Some would take this rather than dairy products.

This one is made especially for ice cream making. It makes the most delicious ice cream that I have ever eaten.

Mr. [John F.] CAREW [D-New York]. How does it go in a punch?

Mr. CARVER. Well, I will show you some punches. [Laughter.] Here is one with orange, and here is one with lemon, and here is one with cherry.

Mr. GARNER. What are these now? I want to learn what these are.

Mr. CAREW. Do these violate the Volstead Law?

Mr. CARVER. No, sir.

Mr. CAREW. Would they?

Mr. CARVER. No, sir. They have nothing absolutely to do with it. If you will allow me to show you one.

I heard some one here ask what kind of box this is. It is a Pandora's box, I guess: it never gets empty.

Here is a bottle of buttermilk. This buttermilk is very rich in fats, and very delightful.

Mr. [Willis C.] HAWLEY [R-Oregon]. Is that made from the peanut?

Mr. CARVER. Made from the peanut milk; yes, sir.

These are derivatives from the buttermilk. That is, the peanut milk has about the same quantity of curds in it as the cow's milk, and leaves this clear curd. It has a distinctive flavor, and you can take it and use any flavoring you want, add any other flavor to it that you desire, and you have your punches and fruit juices, as you wish to call them.

That is another type of punch.

This is the evaporated milk. That is evaporated down much on the order of Borden's milk.

Now here is a very attractive product—an instant coffee. This is instant coffee. All that is necessary is take a teaspoonful of this and stir it into a cup of hot water, and you have your coffee, cream and sugar combined.

Here is a bottle of Worcestershire sauce. That is Worcestershire sauce.

Mr. GARNER. Mr. Chairman, this is very interesting. At least it is to me. I want order in the house.

The CHAIRMAN. Yes. I will have to try to hold the room in order.

Mr. CARVER. This is Worcestershire sauce. This Worcestershire sauce is built on a peanut basis, and you know the original Worcestershire sauce is built on a soy-bean basis, and I find that the peanut makes just as good a base for Worcestershire sauce as the soy bean. So that it comes in for its measure of value in that direction.

Now here is the foundation for the Worcestershire sauce. That is a dry powder, ready for all of the various things that enter into the manufacture of Worcestershire sauce.

Now here is a very highly flavored sauce that imitates the Chinese sauce that enters into chop suey and the various Chinese confections that they are so very fond of.

Here is the dry coffee. Now peanuts make probably one of the finest coffees, of the cereal coffees, that can be made. It is far ahead of Postum. I suppose some of you gentlemen would, judging from what I heard you say awhile ago, look upon it just about the same as a person did at the fair where I had this, and he wanted to know what it tasted like. I told him it was better than any Postum he had ever drunk. He said, "Well, that is no recommendation for it." [Laughter.] Nevertheless, it does make a very fine coffee, imitating the cereal coffee.

Now here is a bottle of curds. Now, as I stated before, the peanut milk has about the same amount of curds that cow's milk has, and the curds can be taken out and made into the various fancy cheeses the Neufchatel and Edam, and any of those soft cheeses, and these curds now are ready to be resoftened and worked into these various fancy cheeses.

Of the oils, we have the oil No. 1, which is the first refined oil, and oil No. 2, which makes a very fine salad oil, and this one was made just before I came up here. I just finished that. This is taken directly from the milk itself, and has properties that the other oils do not have. It is a very beautifully colored oil, and I am looking forward to that with a great deal of interest, as it is a by-product from the manufacture of some of these other things taken, as I stated, directly from the milk.

Mr. CAREW. Did you make all of these products yourself?

Mr. CARVER. Yes, sir. They are made there in the research laboratory. That is what the research laboratory is for.

The sweet potato products now number 107 up to date. I have not finished working with them yet. The peanut products are going to beat the sweet-potato products by far. I have just begun with the peanut. So what is going to come of it why we do not know.

This is the very last thing. Now this is a pomade. That is, it is a face cream and will be attractive to the ladies, of course, because it is just as soft and just as fine as the famous almond cream, and it has the quality of vanishing as soon as put on. It carries a very high percentage of oil and three minutes after it is applied to the skin you can not tell that any has ever been put on at all, yet is a finer softener of the skin than almond cream, and it will take any perfume that one wishes. You can have rose or carnation or any of those fine perfumes.

So therefore the peanut is going to come in as very important in that direction. It is also going to be a fattening element, that is, for the massaging of infants that are anemic and run down. It is going to be quite the equal or above the olive oil because, as I said before, it immediately vanishes into the skin, and is going to be a very attractive thing as soon as physicians find out its value.

Then we have here a bottle of ink. I find that the peanut makes a very fine quality of ink. This is a very interesting thing, indeed.

Then we have the peanut flakes or dried flakes. Now all that is necessary is to dissolve these in hot water, and you have your peanut milk again.

Then we have a relish here. This is relish No. 1, which is a combination of—well, various things. I will not attempt to tell you what it is. But it is a peanut relish.

And then here is a bottle of mock oysters. The peanut curds can be made into mock meat dishes so thoroughly that it is impossible to tell them from meat, and it is going to be very satisfactory in that direction. We are going to use less and less meat just as soon as science touches these various vegetable products, and teaches us how to use them. I remember years and years ago when the automobile first was being introduced how the people laughed and how they jeered and how they talked of the horses, about the impossibility of running them off the streets. I have been here two days, and I have not seen a single horse on your streets. They are automobiles. And now the same thing is true, or much the same thing is true about our vegetable products with reference to the meat business.

Mr. RAINEY. You are going to ruin the live stock business?

Mr. CARVER. No, sir. I don't think I am going to affect it very much, because we will use live stock in another way.

Here is another one of the fancy punches and fruit juices.

Now gentlemen, I have a number of other things. I have probably twenty-five or thirty other things, but if my time is up, I am going to stop.

Mr. GARNER. Let me ask you one question now.

Mr. CARVER. Yes, sir.

Mr. GARNER. I understood you to say that the properties of the peanut combined with the properties of the sweet potato was a balanced ration, and that you could destroy all other vegetable life and continue to sustain the human race?

Mr. CARVER. Yes, sir. Because you can make up the necessary food elements there. Then, as I said before, in addition to that you have your vitamins. You know the war taught us many, many things we did not know before. We did not know anything about these vitamins, and we did not know anything about these various peculiar compounds which are brought out by the complex handling of these various products, and science has touched these things in a way that is bringing to life or bringing to light what was intended should be brought to light, and there is scarcely a vegetable product that we have not learned something about.

Then again, if we think of how the peanut is used, it is the only thing that is universally used among civilized and uncivilized people, and all sorts of animals like it, and I do not know of a single case—that is, I mean normal—that complains because peanuts hurt them. I remember a little boy that we have in our town. Well, he is one of our professor's boys. He made up his Christmas budget, his Santa Claus budget. He started out with peanuts first, and then he would mention a horse, and then peanuts, and then a dog, and then peanuts, and peanuts were the beginning and the ending. He eats peanuts all the time. So that it is a natural diet that was intended that everybody should use. Then again, if you go to the first chapter of Genesis, we can interpret very clearly, I think, what God intended when he said, "Behold, I have given you every herb that bears seed upon the face of the earth, and every tree bearing a seed. To you it shall be meat." That is what he means about it. It shall be meat. There is everything there to strengthen and nourish and keep the body alive and healthy.

The CHAIRMAN. Mr. Carver, what school did you attend?

Mr. CARVER. The last school I attended was the Agricultural College of Iowa—the Iowa Agricultural College. You doubtless remember Mr. Wilson, who served in the cabinet here so long, Secretary James Wilson. He was my instructor for six years.

Mr. [William R.] GREEN [R-Iowa]. What research laboratory do you work in now?

Mr. CARVER. At the Tuskegee Institute, Tuskegee, Ala.

Mr. CAREW. You have rendered the Committee a great service.

Mr. GARNER. I think he is entitled to the thanks of the committee. [Applause.]

The CHAIRMAN. Mr. Carver, if you wish to make any statement in any brief, just file it with the clerk or the reporter and it will be made a part of your remarks. We will be very glad to have you do so.

Mr. GARNER. File it any time within the next week.

The CHAIRMAN. Yes. File it any time within the next week.

Mr. GARNER. Any brief along the same line you have been speaking of here we would be glad to have you file.

Mr. [Allen T.] TREADWAY [R-Massachusetts]. Did the institute send you here or did you come of your own volition?

Mr. CARVER. The United Peanut Association of America, sir, asked me to come.

Mr. TREADWAY. In order to explain to us all this variety of uses of the peanut?

Mr. CARVER. Yes, sir. You have seen, gentlemen, just about half of them. There is just about twice this many more.

Mr. TREADWAY. Well, come again and bring the rest.

The CHAIRMAN. We want to compliment you, sir, on the way you have handled your subject.

Carver's testimony before the House Ways and Means Committee illustrates his remarkable ability to captivate an audience. He handled himself, as President Moton later wrote to him, in a "modest, unassuming manner,"[5] even in the face of obvious racial jibes, such as Mr. Tilson's comment about the watermelon.

Carver's comments also reveal that his research had been given added impetus by the food shortages engendered by World War I. As a scientist, he was committed to using all of nature's resources to improve the standard of living for all Americans, but he was particularly concerned about the South. He envisioned a "New South" that utilized materials and resources peculiar to that region for a greater prosperity. He saw himself, of course, as an important agent of change in that endeavor.

In 1911, for example, he wrote to Booker T. Washington about how he thought native clays could be used to the advantage of poor southern farmers. In this instance, he observed that at least some southerners were already using clay extracts.

June 19, 1911[6]

Mr. B. T. Washington, My dear Mr. Washington:

I think you will be interested in the following observations:

I went out into the country Saturday to make some observations with reference to my work with the clay, and stopped at Mrs. Pugh's. I found that she had a house with four rooms, all of which were whitewashed with the clay, being perfectly white. She tells me that she has used it for years, and has not used a bit of lime. She had some of the clay mixed up in a bucket when I got there, and had just finished using it. She also showed me where she got it. The walls and ceiling of her house were white; the door and door facings were a beautiful slate color. She had made them this color by mixing a little bit of soot with the wash. She had several picture frames painted a bright pretty blue; she had used a white clay and mixed a little bit of laundry blue with it. I also found that others are using it.

On my trip I got some okra that is equal to the beautiful French okra we so highly prize. This was right in the road, and of course I could not investigate very much the extent of the deposit. I also found one of the richest reds that I have ever been able to find.

It really grows more interesting the more I investigate it. I shall make further investigations with reference to the deposits on our own land just as rapidly as I can get to it. I shall be able to show you a number of new colors by the time you return. I feel very sure that none of us except yourself has the slightest idea as to the great variety of uses to which this clay, in its many varieties, can be put, and actually save us dollars and cents. I have one kind out of which I have made a beautiful black color by simply taking some of the clay and mixing a small amount of boiler black with it. By boiler black I mean the soot that comes from our boilers, which is equivalent to lamp black. Now, we purchase a great deal of lamp black, which in many instances this boiler black is just as good. All black used in paint is simply some form of carbon, or, in other words, some form of charcoal, with the proper oils mixed with it. Now, with the great amount of coal we burn here, and the large amount of fine soot that we could collect from our boilers, it would pay us to save this instead of throwing it away as we are now doing. It would certainly keep us from buying the very thing we are throwing away. Just as soon as I get into my laboratory I will make up all these things for you, and shall be able to give you a detailed report.

The more I work with it, the more enthusiastic I am. . . .

Yours very truly,
Geo. W. Carver

In 1922, Mr. J. W. McCrarey, a Phelps Stokes Fellow, wrote to Carver asking a series of questions designed to elicit his assessment of social and economic changes that had occurred and were occurring in the South. Carver responded as follows.

May 2, 1922[7]

My dear Sir:

I take it for granted that you have the answers to the first three questions much better then I can give them to you, so therefore, I am beginning with the 4th and will offer a few suggestions as I have been able to observe them.

IV.

Property owning has been affected:

1st. The people were wedded to cotton; they know nothing else.

2nd. They saw the one and only one great money crop ruined; so to them, the country was ruined and they had no desire to purchase land.

3rd. As yet, no one, two or three money crops have been universally adopted to take the place of cotton.

The average farmer goes on trying to raise cotton in the same old extensive way, which means nothing but failure, more or less, for him. The Country is ruined and he will not stay.

V.

1st. Immigration reduced the enrollment of the churches.

2nd. Those who left contributed nothing, of course; those who remained were financially embarrassed, more or less so the collections fell off greatly. In many instances, a cheaper and more inferior minister had to be employed.

VI. and VII.

Exactly the same applies to five and six.

VIII.

The coming of the Boll weevil in the South and the extraordinary wage inducements in the North and West begat the unusual state of unrest, which meant of course, migration in large numbers.

I do not believe it had anything to do with amalgamation.

IX.

My general feeling is, that as soon as the South can readjust itself to the above condition, that agriculture will be greatly improved on account of more intelligent farmers coming in.

Scientific investigation and demonstration will bring forth other money crops. Factories will come in.

The South, to my mind, will soon become alive as to its vast mineral resources in clays, sand etc.

In time, the whole South and the whole country will be helped.

Geo. W. Carver

When the southern Newspaper Publishers Association celebrated its twenty-fifth anniversary in 1927, Carver took the opportunity not only to congratulate the organization for its long years of service to the South, but also to plug the cause of southern agriculture. This letter was addressed to Marion E. Penn, a member of the organization.

June 16, 1927[8]

Dear Sir:

I beg to extend to the Southern Newspaper Publishers Association greetings on the occasion of the celebration of their twenty-fifth anniversary.

I am sure that few are able to visualize the importance of such a gathering, which is ever on the alert, turning aside here and there studying the great panorama of Southern progress as it passes before it.

We with you are amazed, but convinced that we are not only living in the greatest country in the world, but indeed one of the richest sections of the entire United States on account of its vast undeveloped resources.

Creative and applied science is coming to our relief as never before, and showing us the fabulous wealth we have in our own varied deposits of clay ranging in color from the snow white kaolins and China clays, to the choicest of ochers, rare ambers, vandyke browns, choice Indian reds, beautiful siennas and rare deposits of Fuller's earth.

Nearly one hundred valuable medicinal plants which would yield readily to cultivation, loom up here and there all over the South, making pharmaceutical drug manufacturing plants a paying possibility.

Our equitable climate and responsive soil make the South ideal for trucking, as well as general agriculture.

The cow pea, with its many uses; the soy bean with its thirty-five or more products; the velvet bean furnishing an equal number of products; the pecan with its eighty-five products; the sweet potato yielding its 118 products; and the humble peanut, which some regard as the marvel of marvels, with its 199 products, all emphasize this significant fact that what is true with reference to the products named is equally true with the possibilities for expansion of nearly or quite all of our farm, garden and orchard products.

The rapid growth of industry, the ever increasing population and the imperative need for a more varied, wholesome and nourishing foodstuff makes it all the more necessary to exhaust every means at our command to fill the empty dinner pail, enrich our soils, bring greater wealth and influence to our beautiful South land, which is synonomous to a healthy, happy and contented people.

Yours very truly,
G. W. Carver

In 1930, Carver penned this letter to Mr. Jack Thorington of the Pinckard Investment Company regarding the possibilities for agricultural production and expansion in the South.

February 24, 1930[9]

My esteemed friend Mr. Thorington:

Thank you very much for your letter. . . .

My presence before the Legislature of Texas was certainly something that I was not expecting. I spoke on the possibilities of the South. If creative minds in our institutions could be singled out, set to work along with professors taking chemistry and the dietitians who have to do with the making up of menues for both human beings and animals, that we would move forward this matter of farm relief as never before. The interesting thing and the encouraging thing was that a number of them came up afterwards and said that I was absolutely correct.

Two most interesting editorials appeared in the Dallas News.

I am quite certain that with the constant agitation you are keeping up and the vision you have, we are going to move forward.

I was especially struck with the situation as you outlined it with reference to dairying. Suitable feed for dairy cows can be worked out with our native food stuffs. I hope that science will turn its attention in that direction. We have our splendid grasses, cotton seed and peanut meal, some grains; can have artichokes by-products, which would be comparable to sugar beet pulp. I believe it would pay to grow the artichokes as a cattle food supplement our corn, oats, and other grains. We have our sweet potatoes all of which will work into this great uplifting scheme.

With sincerely good wishes, I am

Yours very sincerely

G. W. Carver, Director

Carver's desire to help the man farthest down was intensified by the suffering he witnessed during the Great Depression. He redoubled his efforts to find solutions to the problems faced by its victims. Having grown somewhat disillusioned with his long-sought fame, at least with his caricature as the "Peanut Man," he returned to his original purpose during the Great Depression, i.e., his work on behalf of the poor African American farmers of the South. This piece, enclosed with a letter he wrote to the editor of *The Peanut Journal* on 15 December 1931, was written in direct response to the Depression. A somewhat revised version appeared in *The Peanut Journal* in January 1932.

Are We Starving in the Midst of Plenty? If So Why?[10]

In Proverbs the thirteenth chapter and the twenty-third verse, we have this statement: "Much food is in the tillage of the poor; but there is that which is destroyed for want of judgement." I doubt if this verse has ever had a greater significance than at the present time.

We have become ninety-nine per cent money mad. The method of living at home modestly and within our income, laying a little by systematically for the proverbial rainy day which is sure to come, can almost be listed among the lost arts.

To illustrate—A few weeks ago, I was visiting in a large city and was entertained in a very luxuriant home of the latest style of architecture furnished with every modern convenience, a Lincoln car of the latest model. The table was furnished with the richest and best in perfect canning and the bakers skill. Yet, when the subject of making a little sacrifice in giving and receiving Christmas presents in favor of the vast hordes of the unemployed, they were not willing to do it and showed very conclusively by their system of reasoning why they needed presents this year more than ever before.

Last summer, we had an unusually large fruit and vegetable crop. Peaches, plums, figs, pears, etc., were often fed to the hogs. Many bushels rotted in the orchards. Any one could get all the fruit they wanted for the asking, yet many families put up absolutely nothing for the winter. Their excuse being too poor to buy jars or cans. It had never occured to them that peaches, apples, plums, pears, figs, cherries, etc, are delicious when properly dried. Corn, stringbeans, okra, tomatoes, pumpkins, etc, all dry easily, kept perfectly and are wholesome and delicious when properly prepared.

Since 1928, the welfare agencies of several sections of our country have sensed the need and have begun to study in a thoroughly scientific and systematic way the whole food problem as it relates to feeding the family, laying special stress upon the food expenditures for low-income families, in order to give them the maximum amount of nourishment at the minimum cost.

With our unprecidented crops of wheat, corn, potatoes, fruits, vegetables; with milk, butter, eggs, etc, etc, and last but by no means least in yield or food value is the billion pound crop of peanuts of good quality. Taking the peanut pound for pound, I know of no other farm, or garden, or field crop that contains as many digestible nutrients pound for pound.

The composition of the peanut appeals with equal force to the habitual meat eater and the vegetarian. The vegetarian sees at a glance that he can get the muscle builders, heat, fat, and energy formers with ease. The lover of meat also recognizes that no other vegetable can be made to give such a distinct meaty flavor to various forms of cookery as the peanut. . . . The ease with which peanuts and its many products blend with almost every kind of food stuff, adding zest, richness, palatability, attractiveness and nourishment, is rapidly taking it out of the confection class alone and making it a necessity in the daily menu, where it rightfully belongs. At this time when it is so important that a low cost menu be made that will contain all the body building nutrients at the minimum cost, within the reach of low-income families, the peanut makes its special appeal.

With milk, butter, cheese, broth, vegetables, etc, etc, the richest most satisfying nourishing soups can be made. With the many fine recipes published in the Journal and other peanut publications, it is easy to find one that will please the most fastideous appetite.

The enterprising and resourceful housewife will be agreeably surprised how perfectly and cheaply she can feed the entire family, giving to each one the maximum amount of nourishment necessary to keep them in good health and strength with some peanuts, milk, butter, vegetables, and skim milk cheese, as the peanut seems especially adapted to these food stuffs, singly or in combination with several.

With these few suggestions, it is hoped that the billion pound peanut crop will be utilized in a way that will bring one hundred per cent nourishment, comfort and joy especially to the many thousand jobless, undernourished people within our borders doing the lions share in keeping the body in fit condition for work as soon as business picks up.

G. W. Carver

Eight months later, Carver enclosed another proposed article for *The Peanut Journal* in a letter to the editor. Again he addressed himself to the need for a creative response to the problems engendered by the Great Depression. This essay appeared almost verbatim in *The Peanut Journal* in November 1932.

Creative Thought and Action One of the Greatest Needs of the South[11]

There have been but few if any periods in the world's history that required more conservative constructive thinking and acting than the chaotic condition through which we are passing.

After spending many months of the most careful observation and analytical study, many great minds agree with Professor Albert Ames, Jr., of the Medical School, Dartmouth College, when he says, "There is but one conclusion to draw and that is that new developments are the life of economic prosperity. If this is so, then, without a steady flow of new developments, prosperity cannot return or be maintained."

Since new developments are the products of a creative mind, we must therefore stimulate and encourage that type of mind in every way possible. The south would in all probability be the greatest beneficiary from the development of these minds by reason of its vast wealth of undeveloped resources, upon which these minds could and would direct their attention.

Never were these old and trite sayings more applicable than at the present time:

(a) Coming events cast their shadows before them.

(b) Strike while the iron is hot.

(c) Take time by the forelock.

(d) The crucial moment.

(e) The psychological moment, etc., etc.

Now is the psychological time for the creative mind to work out the many, many new uses from the inexhaustable deposits of our fine Southern clays; vegetable dye stuffs; mineral deposits, new and old; various and varied mineral waters; Southern fiber plants; material for paper pulp, and many, many other things to numerous to mention in an article of this kind.

One of the greatest needs at the present time is along the line of food stuffs. Secretary of Agriculture, Arthur M. Hyde, urges the return to more normal eating. He says further, that he believes that "Freakish food cults have tended not only to reduce the per capita consumption of food but to divert the diet away from wholesome staple foodstuffs to specialized foods, in the main more expensive but no more healthful or well balanced in nutritional content."

Many and varied are the schemes and menus gotten up to fill the empty dinner pail of the countless thousands who are virtually in bread lines. I have listened to and talked with a large number of persons in charge of these relief measures. I was amazed at the fact that not a single person advised the use of nuts of any kind. The why is easy—information was needed. Who is to blame for the lack of this information? I believe the following persons are more or less guilty:

1. The producer, who feels that he has done his full duty when he depends upon a few time worn advertisements, or none at all, to get rid of his crop. I say get rid of advisedly, as he often sells below the market price by such poor business methods.

2. The merchant who uses only the regular stereotyped methods of display. He puts his nuts in boxes, fancy cartons, jars, bags, etc., and as a rule has no further responsibility.

3. The chemist whose duty ends when he completes and tabulates his long list of analytical data, in terms wholly mystifying to the layman.

4. The dietician completely ignores it in her menus, because none of the books she studied contained anything except a brief reference to the analytical data of a few nuts, and a like number of recipes for candies, confections, etc.

5. The average physician did not know they had any medicinal value because he found no reference to them in the United States Pharmacoposia, so therefore he never got into the picture.

6. The economist did not include them in his various disertations and writings on economy because they are not recognized as a staple article of food.

Now is the time for the application of creative thought in relation to the way it affects the food and commercial value of two of the most outstanding crops of the South—viz the peanut and the pecan.

The pecan is suffering worse than the peanut for new products that will increase consumption. The peanut has to its credit 285 known products. The pecan just a little more than one-third of that number, simply because the creative mind has not taken hold of it in a big way. This great crop with its almost unlimited possibilities will soon be a commercial nonentity unless the creative research workers take hold of it. The pecan could and should be made the King of Nuts, just as the peanut (which is not a nut) is being made the King of legumes. Efforts are being made now to acquaint the consuming public of its great value as a vegetable.

I know of one large peanut factory that has caught this spirit and is making four new products which I consider very choice and a boon to housewives. They are called granulated peanuts at present. Numbers one and two are unroasted. Number one is granulated very fine and is superb for soups, gravies, mock meat loaf, broths, creaming like potatoes, etc., etc. Number two is similar to number one, except coarser—especially valuable for salads and anywhere else requiring a more crispy form of nut.

Numbers three and four are the same as numbers one and two, except the peanuts are roasted before granulating.

I am receiving many letters from persons interested in the use of peanuts in the home, which shows that the housewife is beginning to recognize its superior value as a foodstuff.

As stated before, the pecan (the King of nuts) will grow less and less in commercial value unless the creative research mind is directed to it at once, and a multiplicity of new and attractive uses worked out for the increase of consumption. . . .

The creative research mind should be singled out and given a chance to develop. They would see at once, not only the high food value of nuts as compared with our other standard foodstuffs, but on account of their ease of digestion and with as many appetizing tastes as you have different nuts, together with their attractive appearance, (which is an aid to digestion) emphasizes the great service that could be rendered humanity by training demonstrators, especially in the domestic science departments of our schools and placing their skilled demonstrators at strategic points all over the country to enlighten the people.

The trained home economics demonstrators could be used most effectively I think. I believe also that a strong committee composed of those most affected could work out methods for financing such a movement. No doubt they would seek to:

1. Interest the grower showing that it was to his advantage to make a small donation of nuts to the Home Economics Department of schools with the understanding that they would use them in demonstration work.

2. The home economics demonstrators who work throughout the various counties of the State.

3. The State Legislaturs who, seeing the great good being done would be glad to supplement it by a small appropriation of money from the State.

4. Seek Government aid through our Senators and representatives by bringing such convincing facts to their attention.

5. If necessary appeal to each grower and merchant for a small cash donation of, say, fifty cents or one dollar.

I feel sure that as the marvelous results become manifest, enthusiasm would grow among this group of workers, and it would not be long before nuts would be included in the daily menu where they rightfully belong. One example:

How four people can be served with a generous portion from a few left overs—(delicious): Here is a wholesome, nourishing, appetizing dish costing only a few cents that all will like. Taking one-fourth breast of chicken, the same quantity of cooked ham; run meat through the nash knife of a food chopper; put in pan, season with pepper and salt to taste; add one scant tablespoon of peanut butter, or 1 1/2 tablespoons of finely granulated peanuts; one scant tablespoon butter and one teacup of sweet milk. Boil until it becomes thick similar to fine chicken gravy. Serve over toast, boiled potatoes or hot biscuit.

Note: This recipe can be varied in a number of ways and not lose its food value nor palatability. Any kind of left over meats will do.

G. W. Carver

The *Peanut Journal* essays reveal that while Carver concentrated his attention primarily on the South, he also recognized undeveloped resources in other parts of the country. In 1930, he wrote to a Kansas businessman, an employee of the Mentholatum Company, about the possibility of expanding the uses of wheat.

March 5, 1930[12]

My dear Mr. Hyde:

This is just to extend to you greetings. . . .

I am sure you stand on the only safe and sane ground that will shew us the way out of our educational, industrial, and commercial difficulties. In this connection, I believe that wheat and its low price stands out as one of your challenges to every creative mind, regardless of nationality, complexion, or religious belief. Here lies an opportunity for all to make a lasting contribution to his day and generation. Wheat, one of our oldest grains, dating back almost or quite to the beginning of time, its cultivation is older than the invasion of the shephards, and the Hebrew Scriptures, and yet practically everything we make out of it in a large commercial way is bread, plain or sweet.

The sweet potato has yielded to the creative mind 118 products, the pecan 85, peanut 265. The only reason wheat has not yielded a large number is because the creative scientific mind has not turned its attention in that direction. It can be, and let us hope, will be done.

Again, you are fast becoming an oil center, which means large refineries with their vast and varied output of waste sludges, which can be converted into a number of useful products, experiments now being conducted, show great promise as a road building material, as well as some water proofing substances.

Your state is rich in fine clays; unusually rich in woods, and other forms of vegetable dyestuffs with almost innumerable number of fine medicinal roots, barks, and herbs, which could form a base for large pharmaceutical laboratories. . . .

May this splendid nucleus which you have started spread to other sections until it becomes a world movement and all nations and tongues will be conversing with the Great Creator through the things He has created, interpreting through its many products God's will and wishes with reference to each one.

Very sincerely yours,

G. W. Carver

Six years after this letter was written, Carver wrote to an editor of the *Birmingham* [Alabama] *News* about a plant in Kansas that was producing motor fuel from grain.

October 6, 1936[13]

My dear Mr. Kilpatrick:

"The constant dripping of the water will wear away the hardest stone." This is an age-old truism with tremendous significance. Another, more modern but equally strong in its power, is this: "Constant and well directed agitation will never fail."

Many months ago when the Birmingham News advocated the manufacture of alcohol as a motor fuel from crops from the farm, orchard, and garden, to the less thoughtful it sounded like one of the wildest of "pipe dreams."

On September 25, Atchison, Kansas, boasted of possessing the first plant for the manufacture of power alcohol in America. This plant is using corn. Two batches from the stills have yielded 2,000 gallons, and the officials say that "in a month the capacity will reach 10,000 gallons a day."

The above is most gratifying news. Thanks to the Farm Chemurgic Council for making it possible.

Now that it has become a reality here in the United States, we trust other sections will catch the vision and establish several plants throughout the South at strategic points wherever large quantities of starch and sugar producing crops can be raised.

TAKE CARE OF THE WASTE ON THE FARM AND TURN IT INTO USEFUL CHANNELS should be the slogan of every farmer.

Very truly yours,

G. W. Carver

Carver's reference to the "Farm Chemurgic Council" in the previous letter calls attention to his involvement with the "Chemurgy Movement" that began during the 1930s. This movement sought to discover alternative uses for farm products. In a 1939 book about the chemurgy movement, author Christy

Borth called Carver "The First and Greatest Chemurgist" and devoted an entire chapter to his work.[14]

This excerpt from a 1936 Carver letter to T. Byron Cutchin of *The Peanut Journal* offers a good summary of the ways in which Carver envisioned how farm products could be used to serve the common good.

June 11, 1936

Dear Mr. Cutchin:

. . . . Now is the crucial time to chemicalize the farm. We must not only make the farm support itself, but others as well with a large manufactured surplus to sell to those who are not fortunate enough to own and properly care for a farm. Insulating boards, paints, dyes, industrial alcohol, plastics of various kinds, rugs, mats and cloth from fiber plants, oils gums and waxes, etc., etc., all or most of it can be made from waste products of the farm. Let us hope the creative mind will accept the challenge . . . . [15]

To George Washington Carver, the phrase "waste not, want not" was more than a trite aphorism; it was a philosophical foundation for creative inquiry. He was convinced that there was no such thing as waste in nature—that waste products, or what passed for waste products, were simply undeveloped natural resources. Discovering the many different ways these resources could be put to use, for the betterment of mankind, was the scientist's task.

But, according to Carver, the scientist could not do it by himself. He needed the help of God, the architect and builder of the world of nature, to accomplish his task. It was this last conclusion that set Carver apart from most of the world of science.

FIGURE 1. George Washington Carver as a young boy, about the time he left the Carver farm to attend school in Neosho. (Photo courtesy of the Tuskegee University Archives.)

FIGURE 2. Carver with one of his paintings, early 1890s. (Photo courtesy of the Tuskegee University Archives.)

FIGURE 3. (*Above*) Carver in Tuskegee Institute classroom. (Photo courtesy of the Tuskegee University Archives.)

FIGURE 4. (*Below*) Carver and his colleagues in the agricultural department, Tuskegee Institute. (Photo courtesy of the Tuskegee University Archives.)

FIGURE 5. (*Above left*) In 1906, a decade after Carver arrived at Tuskegee, the institute celebrated its 25th anniversary. (Photo courtesy of the Tuskegee University Archives.)

FIGURE 6. (*Below left*) Carver in his laboratory with students. (Photo courtesy of the Tuskegee University Archives.)

FIGURE 7. (*Below*) Carver at Experiment Station. (Photo courtesy of the Tuskegee University Archives.)

FIGURE 8. Carver traveled widely, often by train, giving lectures and demonstrations of his work. (Photo courtesy of the Tuskegee University Archives.)

FIGURE 9. Dana Johnson, one of Carver's "boys," who corresponded with Carver and visited him at Tuskegee for more than a decade. (Photo courtesy of the George Washington Carver National Monument, National Park Service.)

FIGURE 10. Carver's playful side is evidenced in this image of him pretending to box with a friend. (Photo courtesy of the Tuskegee University Archives.)

FIGURE 11. Carver reading one of the thousands of letters he received from infantile paralysis victims during the 1930s. (Photo courtesy of the Tuskegee University Archives.)

FIGURE 12. Carver first met Henry Ford in 1937. The two men became fast friends and visited each other many times. (Photo courtesy of the Tuskegee University Archives.)

FIGURE 13. A late-life image of George Washington Carver. (Photo courtesy of the Tuskegee University Archives.)

FIGURE 14. Soon after Carver's death, high school teacher Melissa F. Cuther (kneeling, center) helped launch a successful campaign to designate Carver's birthplace a national monument. (Photo courtesy of the George Washington Carver National Monument, National Park Service.)

SEVEN

The Scientist as Mystic

"Reading God out of Nature's Great Book"

More and more as we come closer and closer in touch with nature and its teach-ings are we able to see the Divine and are therefore fitted to interpret correctly the various languages spoken by all forms of nature about us.

Geo. W. Carver 24 February 1930

CARVER'S WORK WITH the peanut made him something of a national curiosity, if not a folk hero. His methods, however, made him suspect among much of the scientific community. Even in his earliest letters, Carver wrote about how he relied on intuition and divine revelation for his scientific insights. Rational thought was for him a way of confirming and illustrating truths that had been attained mystically.

In 1931, Carver wrote to Isabelle Coleman, of Greensboro, North Carolina, of his earliest religious experience and simultaneous conversion to Christianity.

July 24, 1931[1]

My dear Miss Coleman:

Thank you very much for your splendid letter. I thoroughly believe you can get a much better subject to go in your book than myself. After thinking it over again, searching around, if you still feel that I ought to go in there, you have my permission.

The facts in "Upward Climb" are correct, as the writer came here and got the story.

As to being a Christian, please write to Mr. Hardwick, Y.M.C.A., 706, Standard Building, Atlanta, Georgia. The dear boy made a ten days tour with me through Virginia, North Carolina, and Tennessee, where I lectured to a number of colleges and universities. We came together in prayer often to get our spiritual strength renewed. Whenever we come into a great project, we meet and ask God's guidance. Mr. Hardwick will tell you things that I could not. We both believe in Divine guidance. Prov. 3:6; Phil. 4:13; Psalms 119:18, these are our slogan passages.

I was just a mere boy when converted, hardly ten years old. There isn't much of a story to it. God just came into my heart one afternoon while I was alone in the "loft" of our big barn while I was shelling corn to carry to the mill to be ground into meal.

A dear little white boy, one of our neighbors, about my age came by one Saturday morning and in talking and playing he told me he was going to Sunday school tomorrow morning. I was eager to know what a Sunday school was. He said they sang hymns and prayed. I asked him what prayer was and what they said. I do not remember what he said; only remember that as soon as he left I climbed up into the "loft", knelt down by the barrel of corn and prayed as best I could. I do not remember what I said. I only recall that I felt so good that I prayed several times before I quit.

My brother and myself were the only colored children in that neighborhood and of course, we could not go to church or Sunday school, or school of any kind.

That was my simple conversion, and I have tried to keep the faith.

Games were very simple in my boyhood days in the country. Baseball, running, jumping, swimming, and checkers constituted the principal ones. I played all of them.

My favorite song was "Must Jesus Bear the Cross Alone and all the world go free, etc."

If I had leisure time from roaming the woods and fields, I put it in knitting, crocheting, and other forms of fance work.

I am sending you, under separate cover, some literature which may be of service.

Very sincerely yours,
G. W. Carver, Director

In November 1924, Carver, speaking to an audience at the Marble Collegiate Church in New York City, attributed his success as a scientist to divine revelation: "I never have to grope for methods: the method is revealed at the moment I am inspired to create something new." Human learning mattered to him, but it was insufficient; he asserted that "No books ever go in to my laboratory."[2] Two days later, the *New York Times*, in an editorial titled "Men of Science Never Talk that Way," declared that Carver's comments revealed "a complete lack of the scientific spirit" and that they discredited him, his race, and Tuskegee Institute.[3] The prideful Carver responded with a letter to the editor in which he tried to clarify and defend his position, while putting forth his scholarly credentials.

November 24, 1924[4]

My dear Sir:

I have read with much interest your editorial pertaining to myself in the issue of November 20th.

I regret exceedingly that such a gross misunderstanding should arise as to what was meant by "Divine inspiration." Inspiration is never at variance with information; in fact, the more information one has, the greater will be the inspiration.

Paul, the great Scholar, says, Second Timothy 2-15, "Study to show thyself approved unto God, a Workman that needeth not to be ashamed, rightly dividing the word of truth."

Again he says in Galatians 1-12: "For I neither received it of man, neither was I taught it, but by the revelation of Jesus Christ."

Many, many other equally strong passages could be cited, but these two are sufficient to form a base around which to cluster my remarks. In the first verse, I have followed and am yet following the first word of study.

I am a graduate of the Iowa State College of Agriculture and Mechanical Arts, located at Ames, Iowa, taking two degrees in Scientific Agriculture. Did considerable work in Simpson College, Indianola, along the lines of Art, Literature and Music.

In Chemistry, the following persons have been inspiration and guide for study: Justin Von Liebig, Dr. Leroy J. Blinn, Dr. Ira Ramsen, Drs. L. L. De Moninck, E. Dietz, Robert Mallet, William G. Valentin, J. Meritt Matthews, Edwin. E. Slosson, M. Luckiesh, Harrison B. Howe, Charles Whiting Baker, Helen Abbott, Michael, Mad. Currie, Geo. J. Brush, Charles F. Chandler, G. Dragendorff, Frederick Hoffman, Josef Berson, Arthur C. Wright, M. W. O'Brine, Lucien Geschwind, Stillman, Wiley, Dana, Richards & Woodman, Harry Snyder, Coleman and Addyman, Meade, Ostwald, Warrington, Winslow, and a number of others, all of which are in my own library with but few exceptions. In Botany, Loudon, Wood, Coulter, Stevens, Knight, Baily, De Candole, Pammel, Bessey, Chapman, Gray, Goodale, Youmans, Myers, Britton and Brown, Small, and others. These books are also in my own library.

Dietaries, Henry, Richards, Mrs. Potter Palmer, Miles, Wing, Fletcher Berry, Kellogg, Nilson, and others.

In addition to the above, I receive the leading scientific publications. I thoroughly understand that there are scientists to whom the world is merely the result of chemical forces or material electrons. I do not belong to this class. I fully agree with the Rt. Rev. Irving Peake Johnson, D. D., bishop of Colorado in a little pamphlet entitled "Religion and the Supernatural." It is published and distributed by the Trinity Parish of your own city. I defy any one who has an open mind to read this leaflet through and then deny there is such a thing as Divine inspiration.

In evolving new creations, I am wondering of what value a book would be to the creator if he is not a master of analytical work, both qualitative and quantitative. I can see readily his need for the book from which to get his analytical methods. The master analyst needs no book; he is at liberty to take apart and put together substances, compatible or non compatible to suit his own particular taste or fancy.

An Example

While in your beautiful city, I was struck with the large number of Taros and Yautias displayed in many of your markets; they are edible roots imported to this country largely from Trinidad, Porto Rico, China, Dutch Guina, and Peru. Just as soon as I saw these luscious roots, I marveled at the wonderful possibilities for their expansion. Dozens of things came to me while standing there looking at them. I would follow the same or similar lines I have pursued in developing products from the white potato. I know of no one who has ever worked with these roots in this way. I know of no book from which I can get this information, yet, I will have no trouble doing it.

If this is not inspiration and information from a source greater than myself, or greater than any one has wrought up to the present time, kindly tell me what it is.

"And ye shall know the truth and the truth shall make you free." John 8 - 32.

Science is simply the truth about anything.

Yours very truly,
Geo. W. Carver

Early the next year, Carver wrote a letter to a friend, the Rev. Lyman Ward, in response to the incident. Ward, a prominent Alabama white man, was the founder and principal of an industrial school in Camp Hill, Alabama. Carver's letter reveals the pleasure he derived from correspondence supporting his position. He also could not help pointing out that he was in demand as a speaker particularly at white colleges.

1-15-25[5]

My dear Bro. Ward,

Many, many thanks for your letter of Jan. 4th. How it lifted up my very soul, and made me to feel that after all God moves in a mysterious way His wonders to perform.

I did indeed feel very badly for a while, not that the cynical criticism was directed at me, but rather at the religion of Jesus Christ. Dear Bro. I know that my Redeemer liveth.

I believe through the providence of the Almighty it was a good thing. Since the criticism was made I have had dozens of books, papers, periodicals, magazines, personal letters from individuals in all walks of life. Copies of letters to the editor of the Times are bearing me out in my assertion.

One of the prettiest little books comes from Ex. Govt. Osborne of Mich. His thesus on Divine Concord and so many, many dear letters like yours.

I cannot think of filling 1/5 of the applications that are comeing in for talks.

You may be interested to know that the greater part of my work now is among white colleges. I leave this week for N.C. where I will speak at the state univ. state college and two or three other colleges.

Pray for me please that every thing said and done will be to His glory.

I am not interested in science or any thing else that leaves God out of it.

Most sincerely yours.
Geo. W. Carver.

As Carver indicated, the controversy surrounding his New York speech prompted numerous letters of praise. Indeed, as historian Mark D. Hersey has written, this incident "made [Carver] a hero to a Christian culture that by the end of 1925—the year of the Scopes trial—felt increasingly out of step with the larger American culture, perhaps even under attack."[6] Letters of support continued to be sent to Carver. One such letter was written by an Ontario businessman named Robert Johnson, an employee of Chesley Enterprises and a frequent Carver correspondent. Carver responded to Johnson's letter as follows:

March 24, 1925[7]

My dear Brother Johnson:

How very interesting your letter is. I quite agree with you if God did not prompt your letters, you could not write those that really touch the heart as those of yours do.

Of course I can not write such soul-stirring letters as yours, but I will do the best I can. I am so glad you like my motto. I try to live in that way and the Lord has, and is yet, blessing me so abundantly.

Nothing could be more beautiful than your motto Others. Living for others is really the Christ life after all. Oh, the satisfaction, happiness and joy one gets out of it.

I am so interested in the way you manage your saving account. God does indeed arrange it so that it never is quite depleted, unless there is some great emergency, then some soon comes in.

Brother Johnson, I expect to stick to the path, I have no notion of wavering, regardless of how some may sneer. I know that my redeemer lives. Thank God I love humanity; complexion doesn't interest me one single bit. I am not rich in this world's goods, but thank God, like yourself, I have enough to live comfortably.

I have the assurance that God will take care of me. He blesses me with the ability to earn a living, and gives me wisdom and understanding enough to lay a little by from time to time for the proverbial "rainy day."

No, I am not for sale. God has given me what He has in trust to make of it a contribution to the world far greater than money can for myself. Yes my friend, I think I understood you. My letter to you probably was not as clear as it should have been.

I believe that science (truth) if it will take what you have had revealed to you. Search and continue to search. I am sure they will find a world of truth in it.

Less than 150 miles from where I live is one of the unexplained wonders of the country in what is known a "Blue Spring." The pool is about fifty feet in diameter, nearly round. One way you look at it, the water is as blue as indigo. Another way it is as clear as crystal, and you can see down, down, down. In the center is a spot fully four feet in diameter that boils up just as if a huge fire was under it; the water is not hot, not even warmer than other rivers or branch water; in fact, it is a little cooler than the average water. Hundreds of people far and near have examined it.

Large sums of money have been offered to any one who would dive down and find out where the water comes from. Some have tried it so they say, and have gone down a hundred feet or more and had to stop because the water threw them back with its force. No one knows where the water comes from. No difference how much it rains or how dry it gets, this little pool of boiling water is not affected in the least.

Your scripture references, I believe, can well be applied to this case. What I meant was that on the ocean waters journey back to fresh water again, it loses its salt some where, and I believe, in fact, it looks feasible to me that this salt water might travel some distance from the ocean before depositing its salt. If the vein was tapped after the salt had been deposited, a salt mine of dry salt would be the result, if before it gave up its salt, a salt well would be the result.

It seems to me that you have opened a most interesting and valuable line of investigation. . . .

Most sincerely yours.
Geo. W. Carver.

On the same date, Carver wrote to a Seattle minister affirming his own belief in the biblical explanation of creation, despite his adherence to Darwinian thought. In this letter, he seems to affirm the view of man being created in God's image, without endorsing the notion of a short-lived, six-day creation.

March 24, 1925[8]

My dear Rev. Kunzman:-

Thank you for your good letter. . . .

Now as to your question. I regret that I cannot be of much service to you as I have not devoted much time to such investigations in proportion to the almost life time researches of some.

I am interested of course, intensely interested. My life time study of nature in it's many phazes leads me to beleive more strongly than ever in the Biblical account of man's creation as found in Gen. 1:27 "And God created man in his own image, in the image of God created He him; male and female created he them."

Of course sciences through all of the ages have been searching for the so called "missing link" which enables us to interpert man from his very beginning, up to his present high state of civilization.

I am fearful lest our finite researches will be wholly unable to grasp the infinite details of creation, and therefore we lose the great truth of the creation of man.

Yours very truly,
Geo. W. Carver.

This same view was expressed more than a decade later in a letter to the editor of *The Roanoke* [Alabama] *Leader*, the father of a boy whom Carver had treated for paralysis.

March 17, 1937[9]

My estemed friend, Mr. Stevenson:
     Thank you so very much for your splendid letter. . . .
     I attempted to give a little demonstration on the Creation Story as set forth in the Bible and geology. In other words, I attempted to show that there was no conflict between science and religion. I had a great many illustrations from my geological collection, showing many fossils which told there own story. I had quite a large audience, and they seemed to get a little out of it. It was something so distinctly new to them that they probably overrated its value. . . .

Very sincerely and gratefully yours,
G. W. Carver

Carver always sought to merge the worlds of science and religion, particularly in his role as teacher of young people. In 1907, for example, he wrote to Booker T. Washington that he had recently begun a Bible class at Tuskegee. During the next several years—indeed until Carver's death—this Bible class was a Tuskegee fixture.

May 28-'07.[10]

For your information only.
Mr. B. T. Washington.
     About three months ago 6 or 7 persons met in my office one evening and organized a Bible class, and asked me to teach it. I consented to start them off.
     Their idea was to put in the 20 or 25 minutes on Sunday evenings which intervene between supper an chapel service.
     We began at the first of the Bible and attempted to explain the Creation story in the light of natural and revealed religion and geological

truths. Maps, charts plants and geological specimens were used to illustrate the work.

We have had an ave. attendance of 80 and often as high as 114.

Thought these facts would help you in speaking of the religious life of the school.

Very truly
G. W. Carver

Carver's Bible-study class highlights a tension that ran through his religious life. On the one hand, as Henry A. Wallace would rightly recall, "Carver derived more creative nourishment from the Bible" than most scientists of the day.[11] On the other hand, he blended faith in the Bible with a religion that bordered on mysticism. Indeed, he believed God spoke through the natural world as pointedly as through the Bible, and he encouraged his students to find God through Creation no less than through church. He articulated this view clearly in an essay titled "Love of Nature" that he wrote for the December 1912 issue of the publication *The Guide to Nature*.[12]

To me Nature in its varied forms is the little window through which God permits me to commune with him, and to see much of His glory by simply lifting the curtain and looking. I love to think of Nature as wireless telegraph stations through which God speaks to us every day every hour, and every moment of our lives.

No true lover of Nature can "behold the lilies of the field" or "look unto the hills" or study even the microscopic wonders of a stagnant pool of water, and honestly declare himself to be an atheist or an infidel.

The study of Nature is both entertaining and instructive, and is the only sure method that leads up to a clear understanding of the great natural principles which surround every branch of business in which we may engage. Aside from this, it encourages personal investigation, stimulates originality, awakens higher and nobler ideals.

Language fails to adequately express my thoughts regarding the joy of my soul: so, therefore, I send you the above crude paragraphs, hoping that you will find at least one worth of a place in your splendid magazine.

The following letter to Jack Boyd, a YMCA official in Denver, Colorado, explained how Carver hoped young people would come to understand the mysteries of God by studying nature.

March 1, 1927[13]

My beloved Friend, Mr. Boyd:

How good of you to write me, and such a wonderful letter it is.

. . . One of the most beautiful, hopeful, and encouraging things of growing interest, is that there is springing up here and there groups of college bred young men and women, who are willing to know us by permitting themselves to get acquainted with us.

The two little snaps are so beautiful, naturally, God has been so lavish in the display of His handiwork. It is indeed so much more impressive however, when you feel that God met with that fine body of students.

My dear friend, I am so glad that God is using you in such an effective way.

My beloved friend, I do not feel capable of writing a single word of counsel to those dear young people, more than to say that my heart goes out to every one of them, regardless of the fact that I have never seen them and may never do so.

I want them to find Jesus, and make Him a daily, hourly, and momently part of themselves.

O how I want them to get the fullest measure of happiness and success out of life. I want them to see the Great Creator in the smallest and apparently the most insignificant things about them.

How I long for each one to walk and talk with the Great Creator through the things he has created.

How I thank God every day that I can walk and talk with Him. Just last week I was reminded of His omnipotence, majesty and power through a little specimen of mineral sent me for analysis, from Bakersfield, California. I have dissolved it, purified it, made conditions favorable for the formation of crystals, when lo before my very eyes, a beautiful bunch of sea green crystals have formed and alongside of them a bunch of snow white ones.

Marvel of marvels, how I wish I had you in God's little workshop for a while, how your soul would be thrilled and lifted up.

My beloved friend, keep your hand in that of the Master, walk daily by His side, so that you may lead others into the realms of true happiness, where a religion of hate, (which poisons both body and soul) will be unknown, having in its place the "Golden Rule" way, which is the "Jesus Way" of life, will reign supreme. Then, we can walk and talk with Jesus momentarily, because we will be attuned to His will and wishes, thus making the Creation story of the world non-debateable as to its reality.

God, my beloved friend is infinite the highest embodiment of love. We are finite, surrounded and often filled with hate.

We can only understand the infinite as we loose the finite and take on the infinite.

My dear friend, my friendship to you cannot possibly mean what yours does to me. I talk to God through you, you help me to see God through another angle. . . .

Most sincerely yours
G. W. Carver

A similar vision of what he hoped to accomplish with young people is revealed in a 1940 letter to a Tuskegee minister.

August 24, 1940[14]

My esteemed friend, Rev. Haygood:

Thank you very much for your fine letter. . . .

You are quite right with reference to your interpretation about what I mean when I say to young people that I hope they will be bigger than the pulpit. That is really what I mean—that I want them to be bigger than the pulpit and get them to study the great Creator through the things he has created, as I feel that He talks to us through these things that he has created. I know, in my own case, that I get so much consolation a so much information in this way, and indeed the most significant sermons that It has ever been my privilege to learn has been embodied in just that.

I thank you, also, for your sermon at the Greenwood Baptist Church, and if we do not take Christ seriously in our every day life, all is a failure because it is an every day affair. If we can just understand that the Golden Rule way of living is the only correct method, and the only Christ like method, this will settle all of our difficulties that bother us. . . .

Very sincerely and greatefully yours,
G. W. Carver

Carver's correspondence is filled with similar expressions of awe at viewing the mysteries of God revealed in the wonders of the physical world around him. The following letter, for example, written to the Hon. Leon McCord, circuit court judge of the Fifteenth Judicial District, Montgomery, Alabama, illustrates the spiritual feelings evoked in Carver by a particularly beautiful sunset.

December 13, 1927[15]

My esteemed friend Judge McCord:

Yours received yesterday evening.

There are times when ones powers of expression fail to convey the meaning of the heart. I find myself at this moment utterly far adrift upon the high seas without either compass or rudder, as far as the satisfactory power of expression is concerned.

I read and reread your wonderful letter over several times. I reveled in its sublimity of thought and rare literary gift of expression to which the Great Creator has bequeathed to but few men.

There are two things which puzzle me greatly. First, that a person as busy as you must needs be would take the time to write such a letter. Second, and the most puzzleing of all is that you are talking about me, a subject so unworthy of such sublimity of thought and expression.

As I sat in my little "den" reading and pondering over it, nature came to my relief when I was attracted by a strangely mellow light falling upon the paper. I looked up and out of the window toward the setting sun, which was just disappearing behind the horizon leaving a halo of never to be forgotten glory and beauty behind it. It seems as if I have never been conscious of such beauty and sublimity. The variety, brilliancy of color and arrangement were awe inspiring.

As I sat there unconscious of everything except the scene before me, behold, before my very eyes it changed from the marvelous rainbow colors to the soft, etherial "Rembrantian" browns and the midnight blues of Maxfield Parrish. But the most marvelous of all was the pristine light which came from behind those strangely beautiful clouds; the light was like unto bright silver dazzleing in its brightness, and weird in the manner of its diffusion.

As I came to myself I said aloud, O God, I thank Thee for such a direct manifestation of Thy goodness, majesty and power.

I thought of how typical this scene which had just passed into never to be forgotten history was of my good friend Judge Leon McCord, whom I have known for more than a quarter of a century, a person occupying a most responsible and trying position, a position which makes most men cold, severe, unsympathetic, and sometimes cruel, but with my friend, the Judge, many, many thousands will rise up and call him blessed because you have been and are yet ever on the alert to help humanity.

Your "Big Brothers' Bible Class" is one of the strongest testimonials of the above statement.

In this fast approaching season of special reminders of "Peace on earth good will to men", may He who has kept, guided and prospered you during all of these years bring to you and yours additional joys and successes.

Yours with much love and admiration,

G. W. Carver

On another occasion, Carver wrote of how he marveled at God's powers as he observed the beauty of his amaryllis.

3-1-32.[16]

My esteemed friend, Mr. Zissler;-

Before beginning the various routine duties of the day, I feel that I can start the day off in no better way than to pray that all is going well with you, and wish you could share with me the supreme expression of The Great Creator as He speaks to me so vividly through my beautiful Amaryllis (lillies) that are opening daily in my windows in the little den I call my room.

Ten of these great flowers arc open now, one that measures 10 inches in diamater, some of them are striped, spotted and otherwise penciled as exquisitely as orchids.

These are my own breeding and shows what man (in the generic) can do when he allows God to speak through him.

May God ever bless and keep you.

So sincerely yours.

G. W. Carver.

One of Carver's frequent correspondents in the 1920s and 1930s was a young Virginian named Jimmie Hardwick, the same "Mr. Hardwick" to whom he referred in the letter to Isabelle Coleman cited above (for more on Hardwick and "Carver's boys," see Chapter 9). Carver's mysticism comes through frequently in his letters to Hardwick, as in this early effort to encourage his young follower to find God in nature.

7-10-24[17]

Fri. morning. Dear Friend, I feel your loving spirit more than ever this morning. Thank God I feel the growth of the spirit within you.

My Beloved Friend, Mr. Hardwick:

What a joy always comes to me when I recognize your hand-writing in the mail. I always say "bless his heart" meaning a letter from my beloved friend who is more dear to me than any words can express.

I love you and shall continue to do so for the Christ that is in you, both Expressed and unexpressed. I love you also because Christ loves you and longs for you to come into the fullness of his glory.

Your words, my friend, are too strong. There is no danger of your being a hypocrite. You are struggling. You have not lost sight of self yet, but Thank God, you will.

As soon as you begin to read the great and loving God out of all forms of existence He has created, both animate and inanimate, then you will be able to Converse with Him, anywhere, everywhere, and at all times. Oh, what a fullness of joy will come to you. My dear friend, get the significance. God is speaking. "Look unto the hills from whence cometh thy help. Yes, go to the mountains if God so wills it."

Get ready to come down here for a week or so, should God ask you to do so.

Somehow God seems to say to me that this may be so.

For months this vision from times to time comes to me. I think God wants you to begin reading Nature of your own accord first, then when you come here you will learn to interpret it with great rapidity. It may not be here. I may be thrown with you somewhere. Whatever the method is you must learn. Let us pray for guidance.

I have had eight letters from the boys already. Have heard from everyone in the cottage where I stayed. Two of them are coming down soon, they say.

Two have sent their pictures and others are coming. You and those other boys are all wrong. It is not me. I love you because I love Christ in you and whenever you reveal it I cannot help but love it. I loved those boys because Christ was there. As for you, my friend, you belong to me. You are mine. God gave you to me last year. I picked you out of the audience. If I remember correctly while speaking, those great spiritual windows (the eyes) of yours seemed to say, this is the person whom I have chosen to be a great help to you. You need him and maybe you can be of a little service to him.

From that very time until now I have loved you so dearly.

God cannot use you as He wishes until you come into the fullness of his Glory.

Don't get alarmed, friend, when doubts creep in. That is Old Satan. Pray, pray, pray. Neither be cast down or afraid if perchance you seem to wander from the path. This is sure to come to you if you trust too much in self.

Yes, my friend, you are going to grow. Your letters are always such a comfort to me. Do not get away somewhere and fail to write me regularly.

You are now a part of my life and I long for your letters. Well, we both prayed that God would bless the message He sent me to deliver. He really seemed to bless it. At some of the personal interviews the boys wept. I have held my head and wept many times when I read so many of the letters they have written to me.

I fall, my friend, so far short of yours and their rating. God has already told you to go to the Mountains and commune with Him. Why not carry it out without He gives you a new message.

Oh, my friend, I am praying that God will come in and rid you entirely of self so you can go out after souls right, or rather have souls to seek the Christ in you. This is my prayer for you always.

Geo. W. Carver

Some years later, Carver wrote to Jimmie Hardwick about a specific instance in which God had directed him back to a particular research project until he was able to see a significant new truth.

July 1, 1932[18]

My Beloved Spiritual Boy, Mr. Hardwick:

Dear, I must tell you about an experience I had today, which shows so clearly that God moves in a mysterious way His wonders to perform.

Before we left for Miss. while dear Howard was here the first time, I made some collections of fungi some distance away. Occasionally my mind would urge me to go back there again. The urge became so strong this morning that I went. I found the place grown up with weeds and briars.

I began pulling the dead limbs out but the wasps had built a nest in it and soon ran me away without stinging me. I stood afar off quite perplexed, started home, proceeded a little ways and spied another pile of brush. I went to it and found it to be one of the richest finds that I had yet made.

God closed the first door that I might see one open with greater opportunities. This is often so when we are sorely disappointed in some of our findest dreams.

You have seemed to be with me all day today. May God ever be with my great spiritual boy, my pioneer boy, my oldest boy, in fact God's pioneer boy.

With so much love and admiration,

G. W. Carver

Carver's mysticism seemed to increase with his age. He saw purpose and design in every facet of the universe. Indeed, the natural world was his laboratory for discovering the mysteries of an omniscient Creator. This 1940 letter, for example, reveals the thoughts and feelings that were inspired in him when a Tuskegee Institute couple gave him some dahlias. In this letter, Carver writes of the notion of man as God's "co-partner" in nature.

September 7, 1940[19]

My dear Mr. Woods:

This is just to extend to you and Mrs. Woods greetings and to let you know as best I can how much I appreciate the exquisite Dahlias that you brought me.

I remember as a boy a little expression that has lingered with me all through life. It said, "that flowers were the sweetest things that God ever made and forgot to put a soul into it." It was one of the things that impressed me so very much that I always remembered, but as I grow older and study plant life, I am convinced that God didn't forget to do anything that was worthwhile. When we think of the origin of the Dahlia, how it started from a little flower not much larger than a ten cent piece, single only, I appreciate the fact that the great Creator who made man in the likeness of his image to be co-partner with him in creating some of the most beautiful and useful things in the world, and it developed his mind, I can really see why he did not put the soul into the flower. He put it into us, and we expressed it in the development of just such beautiful flowers as you have sent me, and I know that you both are stronger and better from growing these beautiful messengers from the Creator and the fact that you wanted to share them with me is a thought so beautiful that I have no language to express it.

They will last for days and the memory of them, and the spirit which prompted the growing, and the bringing of them to me will always remain.

I am

Sincerely and gratefully yours,
George W. Carver, Director

Even Carver's method of praying was unconventional, as is evidenced by this letter, written late in his life to the Rev. Carl A. Blackman of Kansas City, Missouri.

December 16, 1941[20]

My dear Rev. Blackman:

In answer to your rather difficult request I beg to say as follows:

My prayers seem to be more of an attitude than anything else.

I indulge in very little lip service, but ask the Great Creator silently daily, and often many times per day to permit me to speak to Him through the three great Kingdoms of the world, which He has created, viz.—the animal, mineral and vegetable Kingdoms; their relations to each other, to us, our relations to them and the Great God who made all of us. I ask Him daily and often momently to give me wisdom, understanding and bodily strength to do His will, hence I am asking and receiving all the time.

Very sincerely yours,

G. W. Carver

Perhaps one of the clearest expressions of Carver's belief in the earth as God's workshop appeared in this brief essay, which was enclosed in a 24 February 1930 letter to Hubert W. Pelt of the Phelps Stokes Fund.

How to Search for Truth[21]

I believe the Great Creator of the universe had young people in mind when the following beautiful passages were written:

In the 12th chapter of Job and the 7th & 8th verses, we are urged thus: But ask now the beasts and they shall teach thee; and the fowls of the air, and they shall tell thee.

Or speak to the earth, and it shall teach thee; and the fishes of the sea shall declare unto thee.

In St. John the 8th chapter and 32nd verse, we have this remarkable statement:

And ye shall know the truth and the truth shall make you free.

Were I permitted to paraphrase it, I would put it thus: And you shall know science and science shall set you free, because science is truth.

There is nothing more assuring, more inspiring, or more literally true than the above passages from Holy Writ.

We get closer to God as we get more intimately and understandingly acquainted with the things he has created. I know of nothing more inspiring than that of making discoveries for ones self.

The study of nature is not only entertaining, but instructive and the only true method that leads up to the development of a creative mind and a clear understanding of the great natural principles which surround every branch of business in which we may engage. Aside from this it encourages investigation, stimulates and develops originality in a way that helps the student to find himself more quickly and accurately than any plan yet worked out.

The singing birds, the buzzing bees, the opening flower, and the budding trees, along with other forms of animate and inanimate matter, all have their marvelous creation story to tell each searcher for truth. . . .

We doubt if there is a normal boy or girl in all christendom endowed with the five senses who have not watched with increased interest and profit, the various forms, movements and the gorgeous paintings of the butterfly, many do not know, but will study with increased enthusiasm the striking analogy its life bears to the human soul.

Even the ancient Greeks with their imperfect knowledge of insects recognized this truth, when they gave the same Greek name psyche to the Soul, or the spirit of life, and alike to the butterfly.

They sculptured over the effigy of their dead the figure of a butterfly floating away as it were in his breath. Poets to this day follow the simile.

More and more as we come closer and closer in touch with nature and its teachings are we able to see the Divine and are therefore fitted to interpret correctly the various languages spoken by all forms of nature about us.

From the frail little mushroom, which seems to spring up in a night and perish ere the morning sun sinks to rest in the western horizon, to the giant red woods of the Pacific slope that have stood the storms for centuries and vie with the snow-capped peaks of the loftiest mountains, in their magnificence and grandeur.

First, to me, my dear young friends, nature in its varied forms are the little windows through which God permits me to commune with Him, and to see much of His glory, majesty, and power by simply lifting the curtain and looking in.

Second, I love to think of nature as unlimited broadcasting stations, through which God speaks to us every day, every hour and every moment of our lives, if we will only tune in and remain so.

Third, I am more and more convinced, as I search for truth that no ardent student of nature, can "Behold the lillies of the field"; or "Look unto the hills", or study even the microscopic wonders of a stagnant pool of water, and honestly declare himself to be an Infidel.

To those who already love nature, I need only to say, persue its truths with a new zest, and give to the world the value of the answers to the many questions you have asked the greatest of all teachers—Mother Nature.

To those who have as yet not learned the secret of true happiness, which is the joy of coming into the closest relationship with the Maker and Preserver of all things: begin now to study the little things in your own door yard, going from the known to the nearest related unknown for indeed each new truth brings one nearer to God.

With love and best wishes,

G. W. Carver

One of the more intriguing manifestations of Carver's effort to merge the worlds of science and religion was his struggle in the 1930s to combat the effects of polio and infantile paralysis. Already in the 1920s he was using peanut oil experimentally in massages. When the anemic son of a wealthy white Tuskegee couple gained approximately thirty pounds and became physically active after a series of peanut oil massages, Carver concluded that the nutrients from the oil penetrated the skin and entered the bloodstream.[22] Over the next several years he continued to experiment and often hinted in speeches that he was on the verge of a great medical breakthrough.

In the early 1930s, Carver concentrated his energies on victims of paralysis. Polio was a fearsome disease in those days before Jonas Salk's vaccine, and those fortunate enough to survive it were often left with atrophied muscles and nearly useless limbs. The world was as eager to believe a cure was possible as it was hungry for the cure itself. Consequently, when the Associated Press carried an article in December 1933 describing Carver's self-proclaimed successes in the treatment of paralysis victims, he and his peanut-oil massages became overnight sensations.[23] Would-be beneficiaries of Carver's treatments flocked to his Alabama clinic. Thousands more wrote to solicit his help.

Whatever successes he achieved[24] Carver attributed to God working through him, as he made clear in these two letters to his friend Thomas O. Parish, a Topeka, Kansas, minister.

3-15-35[25]

My great Spiritual friend:-

Your letter came as a great spiritual message to let me know that God was speaking to me through you. Truly God is speaking through these peanut oils that I am working with. Marvelous, some come to me on crutches, canes etc. an in time go away walking. One father brought his dear little afflicted boy 200 miles to me this morning. (Infantile paralysis).

How I do thank you for your prayers. A Ga. pastor of a large church over in Ga. has just informed me that the whole congregation prayed for me last night.

God moves in a mysterious way and is performing wonders. Keep praying for me please.

May God ever bless, keep, guide and prosper you.

Greatfully yours.

G. W. Carver.

September 30, 1935[26]

My esteemed friend, Rev. Parish:

Your splendid letter has been here for a long time. I have been trying to find a few moments that I could call mine in which to answer.

I have now before me 3,000 or more letters from suffering humanity, besides the people who come to see me every day and every night. I often have to refuse to see any one until I can get a little rest.

Your letters to me are great spiritual message. I appreciate your prayers much more than I have words to express.

Ask your congregation to please remember me in their silent prayers that God may continue to manifest through me more of His glory, majesty, and power. I need to become a better medium through which He can speak.

I trust your congregation will learn early the true secret of success, happiness, and power, as embodied in the four passages in the order named:

1. Sec. Cor. 3:5&6
2. Sec. Tim. 2:15
3. Prov. 3:6
4. Phil. 4:13

I am

Most sincerely and gratefully yours,

G. W. Carver

Another of Carver's frequent correspondents in the 1930s was the mother of Jimmie Hardwick, one of his "boys" (see Chapter 9). The following two letters reveal, again, Carver's excitement about the "cure" he had discovered, the intensity of his effort in helping to reduce the pain of what he described as "suffering humanity," and the tremendous workload that the peanut massages placed upon him.

12-16-34[27]

My esteemed friend Mrs. Hardwick:

Thank you so much for your beautiful card with its Greetings. . . .

I have not written to you in a long time, but my thoughts and prayers have been for you daily.

Our patients have usurped almost all of my spare time. God continues to speak through the oils in a truly marvelous way.

I have patients who come to me on crutches, who are now walking 6 miles without tireing, without either crutch or cane. (one man).

My last patient today was one of the sweetest little 5 year old boys, who 3 months ago they had to cary in my room, being paralyzed from the waist down. When I had finished the massage today, much to our astonishment he dressed himself and stood up and walked across the floor without any support.

He is a handsome little fellow and so happy that he is improving. (and I too).

I said Our patients, because I feel that your prayers help to make it possible.

Since last Dec. 31st I have received 2020 letters, plus the people who come every day and almost every night for treatment. It is truly marvelous what God is doing.

Continue to pray for me please that I may be a more fit medium through which He can Speak. . . .

I am so greatfully yours.
G. W. Carver.

3-22-35.[28]

My esteemed friend, Mrs. Hardwick:-

My but it seems so good to hear from you. A person as buisy as you must needs be does not have much time for writing.

How wonderful that you could actually come to Tuskegee, I cannot yet describe my feelings for such a great treat.

I have not been able to make any long trips this fall or winter. I miss being with dear "Jimmie", so much, the precious boy is a real part of my life, but I can understand, why I hear from him so little now, he is on the road.

I hope I can get to the Retreat for one day at last, but I am not sure my strength will hold out. God is surely shewing some of His Glory, majesty and power with some of my patients, they are improving so fast that one is forced to know that the day of miracles are not yet over. . . .

Mrs. Hardwick, I neglected to say the these patients I am working on are our patients, God is answering the prayers of those who are praying for me. . . .

Very sincerely yours.

G. W. Carver.

Carver was always an advocate of prayer, but late in his life he became increasingly convinced that prayer was essential to all that he did, from his scientific experiments to his work with his paralysis patients to his effort to understand the world war that had begun in Europe and Asia. He joined a prayer group led by a Minnesota mystic and religious leader, Dr. Glenn Clark, whose earliest efforts at promoting prayer as a solution to humankind's problems began with his creation of "Camp Farthest Out" during the early 1930s. By the late 1930s and early 1940s, Caver was sending letters such as this one on a regular basis to Dr. Clark. The "little book" to which Carver refers was Clark's biography of Carver, a small book titled *The Man Who Talks With the Flowers*, published in 1939 by Macalester Park Publishing Company of St. Paul, Minnesota.

February 10, 1940[29]

My dear inseparable companion, Dr. Clark:

Your letters, if possible, seem more wonderful than ever as I can see the fulfillment of our greatest hopes and desires from the very beginning. The first crusade we had I felt that it was the beginning of a great spiritual uplift that would slowly but surely sweep the country. It seems to be doing it in such a remarkable way.

Yes, you are indeed with me constantly, and what a pleasant thing it is to have you in the laboratory, in my room, and shaping my thoughts and planning for the future. Dr. Clark is always by my side. No one who has been here has made the impression upon the students, and indeed the teachers as well, as they are yet enthusiastice about your visit and you know how difficult it is to interest students.

The enclosures are marvelous, and especially the one from Cleveland, Ohio. This person belongs to our group. The other one also I wish so much that we had a way of keeping in touch with our various groups. Possibly the scheme you mentioned will do it because it is so significant that we do as it is going to grow more and more as people learn of the

Glenn Clark way which is indeed the Jesus way of living. In fact it is our way. A person came to me yesterday and said in talking about you; "Oh how glad I am to hear from Dr. Clark. His visit here was so helpful, and you know, I believe that he has helped our minister." And I said,—why wouldn't he? Of course he helped our minister just as he helps me. In fact I believe you help me more possibly that anyone else because I know Dr. Clark and the thing for which he stands.

Marvel of marvels, and what a joy it is to have you as a constant companion, one that is truly inseparable, and one that brings peace and joy into my life. My life is filled with joy and peace because I know that my beloved friend Dr. Clark is by my side. I am sure that the Great Creator of all things Who is leading and guiding you will direct your movements in the future as definitely as He has in the past, and that whatever you get out will be just the thing. I am looking forward to it wish so much interest and real enthusiasm. I am convinced, however, that it is impossible for you to even foresee the value of the last booklet you wrote. There isn't a day that passes that I do not get messages from someone who has read the little book. God's little workshop continues to grow in interest.

May God ever bless and keep you. With so much love and best wishes, I am

> So sincerely yours,
> G. W. Carver

Carver's mysticism and spirituality were real and intense. They were the source of his scientific experimentation, but they also caused scholarly criticism of his work and further insecurity in himself. He sought to counterbalance the criticism by redoubling efforts to gain acceptance and notoriety for his accomplishments.

Carver never strayed from the goal of unifying science and religion. In the end, particularly through the medium of his infantile paralysis work and his involvement with Dr. Clark's prayer group, he sought an assessment of his accomplishments in the court of public opinion, especially the opinion of those who shared his view of the value of the "Jesus way of living." Ultimately, he felt, that court vindicated him of the charges leveled at him by "so-called scientists."

EIGHT

Carver

Black Man in White America

I am trying to get our people to see that their color does not hold them back as much as they think.

Geo. W. Carver 27 October 1937

UNTIL CARVER WAS nearly thirty years old, most of the "significant others" in his life were white. His brother Jim was very light-skinned, and there were no other African Americans on Moses Carver's farm as George was growing up. George and Jim lived in the same house as Moses and Susan Carver, the only parents George knew in his early childhood. All of his playmates except Jim were white.

As a child, Carver apparently felt little sense of deprivation or discrimination because of his race. Not, that is, until he and his brother tried to enroll at the all-white school in Diamond, Missouri. Neighborhood parents allowed George and Jim to attend services in their church, but, abetted by Missouri state law, they drew the line when it came to seeing their children being educated alongside *African Americans* (not, of course, the term they used). As a result, the inquisitive, knowledge-seeking Carver left Diamond for Neosho, eight miles away, to attend an all-black school. He was approximately eleven or twelve years old at the time.

He boarded there with Andrew and Mariah Watkins, a black couple who became his second set of surrogate parents. The intuitively bright Carver soon realized that he knew more than his teacher, so when he heard that a white family was heading West to Fort Scott, Kansas, he joined them. He had attended the Neosho school for less than a year.

Late in his life, Carver wrote to Claude J. Bell of Nashville that his first encounter with a large group of blacks occurred while he was in Kansas.

May 15, 1941[1]

My esteemed friend, Mr. Bell:

Thank you very much for your fine letter and the enclosures. The clippings are quite interesting and convincing and the story of Fisk University is equally illuminating.

The first group of colored people that I ever saw in my life was a group of singers from Fisk. I was out in Western Kansas and they came and sang. I was just a mere lad at that time, and the vision and impression has never left me. I thought it was the nearest to heavenly music I had ever heard in all of my life. . . .

Most sincerely yours,
G. W. Carver

At Fort Scott, Carver had one of the most vivid and frightening encounters with racism of his life. In 1879, a young black man accused of molesting a white girl was dragged from the jail and lynched. Carver apparently witnessed some of the violence, as he described it many years later in a brief, undated note to Rackham Holt.[2]

Left Watkins home for Fort Scott, Kansas, with a family who were moving out there. I walked much of the time as they were heavily loaded. Arriving at Fort Scott, Kansas, I had to find a job which did not take me long as I had become more or less skilled in all kinds of homecrafts, such as cooking, laundrying, many kinds of fancy work. I found employment just as a girl. Remained here until they linched a colored man, drug him by our house and dashed his brains out onto the sidewalk. As young as I was, the horror haunted me and does even now. I left Fort Scott and went to Olathe, Kansas.

Later, as was mentioned earlier, Carver was denied admission to Highland College in Kansas, after school officials there discovered that he was African American. And then there was the overt racial hostility that he experienced during his early days at Iowa Agricultural College.

Another incident that drastically affected Carver's reaction to white society occurred in Ramer, Alabama, just a short distance from Tuskegee, in neighboring Montgomery County. The cause of the incident was the appearance in the town of a white photographer, Frances B. Johnston, with a black teacher, Nelson E. Henry. Miss Johnston was traveling throughout the South, gathering information on black schools. Carver described what happened in a letter to Booker T. Washington.

November 28, 1902[3]

Dear Mr. Washington:

I have just returned from a trip to Ramer, Alabama, where Mr. Henry is located, and I feel that you ought to know the exact condition there as it is the most distressing of any that I have ever seen in any place. In fact, I had the most frightful experience of my life there and for one day and night it was a very serious question indeed as to whether I would return to Tuskegee alive or not as the people were thoroughly bent upon blood-shed. In all probability they have broken up the school. Mr. Henry was obliged to leave Ramer between the suns and the other teacher became so very frightened that she left also. The occasion of the disturbance was Miss Johnston, who went down on the same train that carried me down. The white people evidently knew that she was coming. The train was late in getting there but a number of people had gathered at the station to see what would happen. I took Miss Johnston's valise and put it in the buggy for her. Mr. Henry drove her to his house and put out her valise and started to the hotel, then he was met by parties and after a few words was shot at three times. Of course, he ran and got out of the way and Miss Johnston came to the house where I was. I got out at once and succeeded in getting her to the next station where she took the train the next morning. The next day everything was in a state of turbulency and a mob had been formed to locate Mr. Henry and deal with him. They did not pay a great deal of attention to me as I kept out of the way as much as possible, but it was one of the worst situations that I have ever been in.

As things are now, the school is broken up and there seems to be no way of settling the difficulty. They say that what they want is to get hold of Mr. Henry and beat him nearly to death. I spoke to the people on Wednesday and they were—of course—very much disturbed. I quieted them down as much as I could—which was very little. I had to walk nearly all night Tuesday night to keep out of their reach. Wednesday night I stayed four miles from the place and took the train six miles from Ramer next morning. On Wednesday the place was patrolled by a white man walking up and down in front of the school house armed with a shot gun. I went down on Wednesday morning to see just what the situation was and I saw twelve horses saddled and tied to the fence of one of the chief promoters. He saw me coming down the railroad and at once mounted his horse and came down to meet me. I stopped aside to examine some plants—just to see what he would do—and he came up and eyed me closely and spoke rather politely. He evidently thought

I was Mr. Henry. One of the gentlemen went down town that night to see what was being done and found that a mob was being made up for the night to take Mr. Henry. I succeeded in getting word to Mr. Henry to flee for his life, which he did. He is now in Montgomery.

Mr. Henry gone, they then telephoned over to the next station to see if Miss Johnston took the train next morning. They wanted to know who took her to the train, and everything in detail. A telegram was handed to a gentleman, which was evidently a fake—at least appeared so. It was purported to have come for Mr. Henry to induce him to come to the station. It was simply to find out where he was. I have never seen people so enraged.

Mr. Henry was doing a great work there and it grieves me to know that he must give it up. Miss Johnston was thoroughly grieved. I might say that she is the pluckiest woman I ever saw. She was not afraid for herself but shed bitter tears for Mr. Henry and for the school which is in all probability broken up. They were preparing to have a splendid exhibition. In fact, the material was there and promised to be one of the best exhibitions that I have had the privilege of attending. The exhibits were large and fine and the people seemed very much encouraged. Now as to the outcome, it is impossible to say. It stands just as I have related it to you. Mrs. Washington and I have talked the matter over here and we think it wise to say just as little as possible about it here. The people seem to be intensely bitter against any one who comes from Tuskegee.

Trusting you are quite well and that you had a pleasant Thanksgiving, I beg to remain,

Yours most sincerely,
G. W. Carver

These experiences, combined with less dramatic encounters with white racism, left Carver with conflicting, even confusing, views of whites. On the one hand, it was whites who raised him and helped him, and even loved him. On the other hand, he saw how irrational and vicious—even deadly—white wrath could be. A gentle, accommodating man by nature and nurture, Carver resolved to be even more accommodating and cautious in the future. He viewed human beings of all races as members of a God-created family and was troubled by whites who declined to share that perspective. He was equally impatient with blacks who, in his judgment, contributed to racial stereotypes by living the low life. One of his most elaborate written statements on the subject came in a letter to his long-time friends, Mr. and Mrs. John Milholland. The Milhollands had apparently written to Carver, with some distress, about prejudicial attitudes held by their own family members. Unfortunately, the letter to which Carver was responding is not extant.

Feb. 28 - 1905[4]

My dear Mr. & Mrs. Milholland,

. . . I am so glad to hear you say what you do about the race question. I can never think of you and your family ever taking any other stand than that a "man's a man" ect.

How could Mrs. Olive be changed she always seemed so good and true.

For eight long years I have labored here, and oh so often I have been shocked and made sick at heart over the many terrible things perpetrated upon our brother and sister man on account of their color.

The Southern people seem to have a way of working over those who would be our friends under more favorable circumstances.

Yes it is the "diporats" ["depot rats"], "livery stable", gangs and the general worthless class she has constantly come in contact with. The jentle, refined, cultured, self sacrificing negroes she sees but little of because they do not make themselves conspicuous.

And as you say, the finer examples, we are a young race yet, not by any means perfect but every day and year marks a part or complete milestone upward.

Oh how I pray that the light may burst forth in all of its splendor upon such unfortunate people.

Slavery was a hard and terrible school but I am sure it had something good about it despite the fact that many a heart was broken, and many sank into eternal rest to open their eyes into another world and reap their reward, whatever that was.

You know full well that we see in people and animate things just about what we are looking for, How I wish Dr's nephew would look for the good as well as the bad. I am sure he would find it.

I am confident that God in his own good time will bring the right things to pass.

I am so surprised that you have those crude remembrances of me, how I wish that I could see you all I so often think of you. . . .

With very best wishes
Geo. W. Carver

Carver endorsed Booker T. Washington's accommodationist "solution" to the race problem, as did many southern blacks of his generation. Early in his tenure at Tuskegee, Carver included the following thoughts about Washington's more militant black critics.

8-27-98[5]

Mr. Washington I hope you will not let any such articles similar to that of Paul L. Dunbar give you a moments uneasiness but simply stimulate you to press on. You have the only true solution to this great race problem. It is only ignorance mostly and a bit of prejudice that prompts such articles. Among both white & black, you are living several hundred years ahead of the common herd of both races. Many of our own dear teachers here are just as blind as can be only live in the present 3 or 4 hundred years from now people will know and honor your greatness much more than now because they will have been educated up to it. Pardon me for taking so much of your val. time May God bless you & Mrs. Wash.

Press on.

Very Respy.
Geo. W. Carver

Years later, long after Washington's death, Carver paid Tuskegee president Robert Russa Moton what he considered to be the supreme compliment when he compared Moton's ability to get along with the white community favorably to Washington's.

April 13, 1927[6]

My dear Dr. Moton:

I think you ought to know the attitude of some of the leading white people up town. They are regarding you now as never before with the same kind of cooperative spirit accorded to Dr. Washington.

They feel very happy over your speech made out in the country when the Trustees were here. I have heard some of them quote some of the things you said several times. Each laid emphasis upon you as a disciple of good will and friendly relations. . . .

Very sincerely yours
G. W. Carver

In 1923, Carver was the recipient of the prestigious Spingarn Medal, given annually by the NAACP to the African American considered to have made the greatest contribution to the advancement of the race during the previous year. No doubt the publicity surrounding his testimony before the House Ways and Means Committee in 1921 had helped him tremendously. Carver was elated over the honor, particularly because it resulted in greater recognition and praise among whites. He wrote the following letter to the man who created the award, Joel Spingarn, a white publisher and former chairman of the board of the NAACP, a board that, incidentally, included W. E. B. DuBois, whose solution to the contemporary race problem differed dramatically from that of Carver and Booker T. Washington.

October 8, 1923[7]

My dear Major Spingarn

How delighted I was to get your good letter. . . .

The larger view as to the medal's value, to my mind, supersedes everything else and is giving a kind of education that nothing else will give. It is certainly having its effects right in our own little town. The white people seem to be even more anxious to see this medal than my own people. I must confess again that I do not feel worthy of such distinction. However, I shall endeavor, with all that is within me and as fast as the great Creator gives me might, to at least make my friends have no regret that it came this way.

I am working on some distinctly new problems now and they seem exceedingly promising. One of the things that has interested me possibly as much as any other is the stimulus that my work is giving to real investigation. On the strength of this work, a corporation known as the Industrial Waste Products Corporation has sprung into existence. They say that their work was suggested by the investigation made by me. I am sure that scientific thought is going to be turned more and more in this direction. Schools and colleges will take it up.

I have just received a letter that interests me, a portion of which I will quote: "The students who were at the Student Conference held at Blue Ridge last summer are beginning to inquire as to whether or not you could come to some of the various colleges during the coming term. Some of them may write you directly. It has occurred to me that if you would convert your time in visiting all of these colleges at one time, it would make your work more effective in the white colleges. I have before me a request for at least three colleges in South Carolina: Wooford, Furman and Clemson. There will no doubt be others. I am writing to know if at any time during the term, it would be possible for you to spend a week speaking in these white colleges in South Carolina? If so, what dates would be most convenient? I would hope that you might have in addition to addressing the students, a day for an exhibit so there might be an opportunity for questions and discussion with them personally."

This letter comes from Mr. Will W. Alexander, Director of the Commission on Interracial Co-operation, with headquarters at 409 Palmer Building, Atlanta, Georgia.

With sincerely good wishes, and again thanking you most heartily, I am

Very truly yours,
G. W. Carver

Meanwhile, Carver continued to praise President Moton, emphasizing what he believed to be Moton's contribution to improved relationships between the races

July 12-28[8]

My dear Dr. Moton:-

This is simply to extend to you greetings, and to say that I have been reading with so much interest, delight and satisfaction your various activities this summer.

My chief anxiety is for you, Dr. Moton. I fear you are not getting much rest.

You are making such valuable history for the race along so many lines that it would be a jenuine calamity to have it curtailed in any way.

To my mind you are teaching the southern white man to know us in a way that has never dawned upon him before.

More young people (the hope of better race relations) are coming here and really studying the Negro than ever before.

Take some real rest Dr. Moton. May God ever bless, keep and preserve you.

Very sincerely yours,
G. W. Carver

Carver's support of an "accommodationist" approach to race relations did not mean, however, that he was without racial pride. When a representative of the Peanut Growers' Association asked his opinion about an advertising campaign that included references to "Pickaninny" peanuts, he voiced his objections, albeit in what he hoped would be an inoffensive way.

November 29, 1929[9]

Dear Sir:

I beg to acknowledge receipt of your interesting favor of recent date. . . .

I believe your idea is a good one and with the advertisement you have in mind, I do not see why the venture should not become popular. I trust that you will, however, understand me and pardon me for making these suggestions.

It is solely in the interest of the peanut industry. I notice your trademark is "Pickaninny" peanuts, and that you are going to have a particularly appealing "Pickaninny" face.

I take it for granted that you are putting up these peanuts with the hope that everyone will buy them and that your trademark will become very popular. Now, my people object seriously to their children being called "Pickaninnies" as the usual Negro child. "Pickaninnies" as they are called by some, are merely caricatures. . . . I presume that you are acquainted with the unpopularity of the "Pickaninny" pie which was made after the manner of the Eskimo pie, but the caricature of its trade-mark made it very unpopular, so much so, that I understand that the originator had to give up the business. Of course, the "Gold Dust Twins" washing powder, and the "Cream of Wheat", both colored advertisements, are very popular.

Now, as I stated before, I trust that you will understand me in what I mean and in using these trade-marks, do not have it an ugly cartoon. . . .

I shall be glad to co-operate with you in any way I can, with sincerely good wishes, I am

Very truly yours,
G. W. Carver

Another manifestation of Carver's pride in his race was that he sought to publicize the accomplishments of African Americans. He wanted black and white youths, in particular, to be aware of the creative genius and accomplishments of "colored people." That was the message contained in a letter to Carter G. Woodson, founder of the Association for the Study of Negro Life and History and editor of the *Journal of Negro History*.

May 10, 1929[10]

My dear Mr. Woodson:

Please send me, at your earliest convenience, a list of the pictures, books, etc., that you handle with reference to colored people.

I have just returned from Tulsa and other points in Oklahoma. I dedicated a large Junior High School at Tulsa, and the question of books with stories about Negroes who have made good was brought up, as the Superintendent of Education desired to put them into the colored schools. I told him, as best I could, about your publications. I had a copy of the pictures you handle, but it has evidently been given to some one.

I believe that there is a great opportunity along this line.

I shall be going out again to some very important schools within a few days, and I should like very much to get this information as early as possible, and I believe I can render your work a very distinct service.

Very truly yours,
G. W. Carver

Carver enjoyed speaking to white audiences. In a land where he had to travel, sleep, and eat as a second-class citizen, he relished the opportunity to captivate crowds composed of a race that considered African Americans as inferior. It is obvious from the following letter to Jasper, Alabama, school principal Albert J. Taylor that he enjoyed the idea of whites being "stirred up" by his speeches.

February 8, 1928[11]

My dear Mr. Taylor:

I am very glad indeed to get your good letter which I found awaiting my return. I have studied it with considerable care, as I wanted to see whether there is a possibility of my coming back sooner.

My schedule at present as you may well know is very much crowded, and I cannot say definitely just when I can make the trip again. I have some functions coming off here which will demand my presence. The dates have not been definitely set. Then, I have quite a lengthy tour to make through Virginia. The dates are being worked upon now. There are also a number of local demands which I shall have to meet. I shall endeavor to keep in touch with you from time to time as I did before.

I have received two long, interesting letters from Mr. Moore, Superintendent of the Board of Education. One just came this morning, and he says that the white people are very much stirred up and want me to return. He is going to send me some material for investigation. It stirred up so much interest that I regret that we did not have it in the Court House, or some large auditorium where more could come. I shall do the very best that I can to come again. I received a copy of the Birmingham

Truth and enjoyed reading the write-up very much. Very excellently done. If any other papers carry it, I shall be very glad to have them.

I certainly remember with much pleasure and satisfaction my trip.

Trusting that I may keep in touch with you from time to time, I am

Yours most sincerely,
G. W. Carver

Carver believed, of course, that the significance of his work transcended racial boundaries. His goal was to educate the entire human race to the "vision" he had of using the gifts of nature for the betterment of mankind. He elaborated on this thought in a letter to J. Alex Moore, the Walker County, Alabama, superintendent of the board of education, who was mentioned in the above letter.

February 3, 1928[12]

Honorable Sir:

This is just to extend to you greetings and to thank you for your good letter which is appreciated far more than my words can express.

Two feelings came over me while in your beautiful little city. I was struck first with the broad open minded, progressive body of people you have there; indeed it came not wholly as a surprise, as this is just what I expected with such a man as yourself as Superintendent of the Board of Education.

Every since I received your first letter, I have cherished the possibility of coming to Jasper and meeting a large number of young people, of both races, because the work which God has given me to do as a trail blazer, I believe is a distinct contribution to education.

I must go the way of all the world before long, and it is so important that our young people catch the vision.

Mr. Moore, I regard this work as not a racial problem at all, it is too big for that. It is a great human problem destined to affect, sooner or later the whole civilized world, because what is true of the sweet potato, peanut, velvet bean, soy bean, cowpea, pecan, clays, fiber plants, and many, many other problems with which I have worked and am working, is more or less true with every other farm, garden, orchard and miscellaneous products.

All we need, Mr. Moore, is to set the creative mind to work. In this work I am not interested in complexion or nationality, as each country, state and county has its own individual problems, to some extent, which will fit into a great educational scheme, when enough well trained, creative minds can be set to work.

There need be no collision in the status of the workers, each would work out his visions in his own way, just as fast as God gives him (in the generic) light and strength.

I love humanity, I am interested in young people, for they must catch the vision.

In this I was disappointed that so few were present.

I too cherish the hope that it will be possible for me to come again. Should this be so, I will not come until I am sure you have been consulted as to proper arrangements.

I am not in the slightest way critizing what was done, because I was well taken care of and suffered no ill effects from the trip. I regret, however, that my physical strength will not permit me to do much traveling now.

I have so many things under investigation here now. Marvelous indeed, some of them are. You would be greatly surprised if you could be here a day or so and see the immense number of people who come to the laboratory daily for consultation. While I was away some people came from New York City to see me, waited the entire day but had to leave.

Please pardon this long letter, but I just wanted you to know a few of the things that are ever uppermost in my mind. I may be all wrong, if so I shall be very happy to have you set me right.

The meeting with you and the delightful group you had at the meeting was a very distinct pleasure.

How I wish I were just half of what you said in your introduction and your fine letter. It places a new responsibility upon me to try to live at least within hailing distance of it.

I am

Very greatfully yours
G. W. Carver

The following letter to Dr. G. F. Peabody provides several interesting insights into Carver's attitude toward what he described as "the so-called Negro problem." Carver indicates his belief that human progress is "practically inherent within the individual," that lack of African American initiative, in addition to white racism and oppression, is the cause of black backwardness. Carver also indicates that "carefully worked out control measures" should be used in urban areas where massive black migration (the so-called "Great Migration") was causing racial strife. He suggests that a much better solution would be to keep blacks in the South by providing them employment opportunities through the medium of scientific agriculture of the type being propagated at the time by businesses such as his own Carvoline Company.

September 20, 1923[13]

My dear Mr. Peabody:

How happy I was to get your good letter which I have read many times, and a real treasure which I prize much more than my words can express. I do not prize it simply because you have said such fine things about me personally. I am not, however, lacking in appreciation, as I am fully conscious that words coming from such a high authority as you, are enough to make one feel justly proud.

I feel, however, quite unworthy of such a beautiful letter. I am simply trying as best I can and as fast as God gives me light to do the job I believe He has given me in trust to do. . . . Rising or falling, I believe is practically inherent within the individual, and since races and nations are made up of individuals, they progress or are held back by the percentage of individuals who will, or will not to do the right thing.

I have lived in the North, west and South and I have been pleasingly gratified at the large number of people who will really go out of their way to see to it that those who are inclined to do the right thing get a square deal. Divine love acts upon humanity the same everywhere, and Divine love is going to rule the world.

I pray daily that I may prove worthy of every confidence placed in me, and never loose faith in humanity, or my love for humanity. Your life has helped me greatly. I believe in the providence of God working in the hearts of men, and that the so-called, Negro problem will be satisfactorily solved in His own good time, and in His own way.

I am not surprised at stupid and ignorant out-breaks. It seems perfectly natural to me that these people are congested in large numbers, subjected to an environment, wholly unlike any they have ever experienced before, and naturally many of them are not able to properly adjust themselves to their new conditions. Hence, they go off on a tangent.

I believe that the city or town where large numbers of these people congregate should institute carefully worked out control measures, such as would help them to properly adjust themselves to their new and strange environments. By so doing, the unfortunate outbreaks and racial troubles could be reduced to the minimum, and finally obliterated altogether. I think the influx of a large number of white citizens as a core, is neither possible, probable or desirable.

I believe more strongly now than ever before that the south is the richest section of the whole United States, on account of the vast number of undeveloped resources, and I hope in the near future, we will become a great manufacturing section, as well as an improved agricultural, dairying, and stock-raising section. Many of these people would remain in the South and make fine factory laborers, but now we have no factories for them to go into. A more intelligent agriculture is going to be brought about.

The work that I am trying to do has the above in mind and its goal. The new company seems to see this. I sent you their prospectus the other day, which I think will give you the information you desire.

I trust you will pardon this rather long rambling letter.

Sincerely yours,
Geo. W. Carver

In that same year, Carver spoke at a white Atlanta school. He was overwhelmed by the reception he received from whites, as this letter to Lyman Ward, principal of the Southern Industrial School at Camp Hill, Alabama, makes clear. No doubt the special favor whites showed him encouraged him to believe in the merits of accomodationism.

April 23, 1923[14]

My esteemed friend, Mr. Ward:

I wish to thank you for the wonderful write-up that you sent me. This is what I have been hoping to see for some years. I trust that it may be given wide publicity. It is very fine. It does me so much good to read it and it was so good of you to send it to me.

It may be of interest to you to know that a special car came from Atlanta on the 6th of this month, remained at the school twenty-four hours (24) waiting for me to get ready, took me to Atlanta, and remained there until the 13th. I lived on the car during a stay there. I think this is the first time in the history of the Negro race that such has happened. I cannot help exclaim, "what has God wrought?". Throughout the entire visit, nothing but courtesies were extended. I thought you would be interested to know this.

The purpose of these visits is to exploit the clays, sweet potatoes, peanuts, pecans, and other products that have been worked out here.

The organization seems to be moving along nicely. Those who have given it careful study seem to feel that it will revolutionize economic conditions here in the South.

Again I wish to thank you for sending me the fine write-up of your Institution.

Yours most sincerely,
G. W. Carver

Carver was, at bottom, an optimist with regard to the South's ability to transcend animosities and reform itself into a land where people could be accepted without regard to race. Such was the message of this 1929 letter to the Christian socialist Howard A. Kester, a long-time Carver friend, head of the Youth Section of the Fellowship of Reconciliation (New York), and a future co-founder of the Southern Tenant Farmers' Union.

April 2, 1929[15]

My dear Mr. Kester:

I have read with much interest and satisfaction your plea for an office in the South for the establishment of better race relations.

The surprise to me is that this has not been done before now.

The beating on the tail of the snake may stop his progress a little, but the more vital parts must be struck before his poisonous death dealing venom will be wiped out. Just so with the poisonous venom of prejudice and race hatred.

I believe the Southern people will welcome such an office, and many will lend their cooperative support.

May I close with this statement from one of our most highly educated Southern divines, which expresses the growing feeling of the South with reference to race justice and fairness more eloquently than I can put it.

He had this to say through the columns of our country paper, February 7, 1929.

Brothers All of Us.

"We are brothers, all of us, no matter of what race or color or condition; children of the same Heavenly Father. We rise together or we fall together.

No one can triumph or fail alone. If one suffers the other must suffer. If one succeeds, the other must rejoice. We are members one of another, one body.

Let us not forget the head of that body is Christ."

Where there is such a growing sentiment as this, right in the heart of the great conflict, it seems to me that untold good could be done if such an office as you propose was located in the South, where it could render at least first aid.

Yours very sincerely,
G. W. Carver

One of Carver's solutions to the "race problem" was the emergence of an elite, educated, creative black populous who could serve as models of achievement to the black masses and as evidence of black potential to the white citizenry. Ironically, this attitude bore at least some similarity to the "talented tenth" philosophy of Booker T. Washington's old nemesis, W. E. B. DuBois.

Carver, of course, saw himself as this kind of model, and he had a friend, Dr. M. L. Ross, a Topeka, Kansas, physician and fellow experimenter, in whom he saw the same qualities. The following series of letters contains numerous expressions of Carver's belief in Ross's ability to inspire.

July 27-30.[16]

My very dear friend, Dr. Ross:-

Your splendid letter has just reached me and brought such good news. . . .

As far as you are concerned I mean more than I say. You are moving in the direction that will do more to solve the race, so called, inferiority complex than all the speeches and beautiful oratory that could be made in a lifetime.

What the race needs is creative powers and less imitative.

I have no fears concerning you measureing up to my expectations you are going very fast in the right direction.

Of course you are satisfied to do hard work, it becomes a pleasure, you do not have to wait so long for results. You have gotten some splendid ones already.

Your brunette powder seems excellent. I presume you have tested it out to see that there are no fine abrasives or chemical substances that will injure the most delicate skin, be sure you test it out most carefully fore these.

I am delighted with its color and texture.

You, my friend, have hit the key note, the world is full of useless products, products that are incomplete, the mad rush for gold has filled the country with all sorts of "get rich quick" schemes and products, many good products have been ruined because they have not been given the time test which alone can bring out the true merits of a product.

In the short time you have been doing this work, it is impossible to tell how many young minds you have inspired.

Your face powder will be fine for your exhibit at the fair this fall.

You are trying Penol out in a new way. I never heard of it being used for back complaint. I shall be most interested to know further reactions.

I beleive you are quite right, peanut milk is an exceedingly complex emulsion containing, all of those high, and rare protied bodies of which we know little about in addition it contains some Alkaloidal properties, which are strictly medicinal. . . .

With much love and best wishes to you and Mrs. Ross.

I am admiringly yours,

G. W. Carver

Sept. 15 30.[17]

My beloved friend, Dr. Ross:-

Your letters are all such treasures, I keep every one of them, they have only one fault and I am unable to get it corrected and I have just made up my mind to grin and bear it.

Your over estimation of me 75% off is the fault.

If I could get (and I say this in all seriousness) a few more men out of the race like Dr. Ross, who talked but little and did much what a blessing it would be. Just think of it, if we had just one man like you in every community, it would transform the race in the eyes of the world, because we could become indispensible factors in the development of this great country in which we live.

Dr. Ross, I really feel very proud of you and what you have done and are doing. . . .

Admiringly yours,

G. W. Carver

Oct. 1-30.[18]

My beloved friend Dr. Ross:-

Your wonderful letter reached me today and it not only cheered me but I am thrilled at your vision. . . .

All that I have said and will say about you is that I am perfectly satisfied with you ambition and ability, also the way you go at things.

O, how I long to see "an" Dr. M. L. Ross in every community, there would soon be no aggravating race problem to my mind. . . .

There is no question but what you are going to do something worth while.

The way you inspire young people to do their best is a real worth while contribution.

I trust you will cary out your ideas with reference to the clay.

You will neccessarily have to go slow with all of your work as you will over do your physical strength. . . .

I too hope you can see my laboratory some day. I want you to know how thoroughly I appreciate your kind thoughts of me. I am now visiting cotton farms also white people come and get me to look over their crop, are very tender and considerate of me, so that I beleive it is doing much toward breaking down the color barriers and making it easier for the dear young people of my race who must carry on long after I have been called hence. . . .

Yours with great admiration,
G. W. Carver

Early the next year, Carver told Ross that he often used the latter as an example of "colored people with creative minds."

Jan. 21-31.[19]

My beloved friend, Dr. Ross:-

. . . My dear friend. I felt that your Creative mind and soul would thrill at the little card I sent you as it would offer you suggestions.

You too are in my mind equally as many times. People ask me almost daily, have you or do you find colored people with creative minds. I say Yes, and mention you as one who is actually doing things. I am happy to have you tell me that you are well again and that you are going to begin on your Creative work and let the children have the benefit of it.

You will do untold good. All the Creative mind needs is inspiration to go ahead it will find ways of doing it.

My friend, I can imagine just how far behind your work is, but take it slowly and cautiously.

I hope you will arrange to come before the children and others, Just as soon, and just as often as possible with your creative work.

O yes God lets us go one step at a time and as fast as people can interpret our meaning, or in other words live up to the light God has given us. We are ready to explore larger fields, and leave our work behind for those who are to come after us. This is as it should be.

I too hope and some how believe that our paths will meet in this year. I confess to you that I grow just a bit chesty when I think, that God directed me to your beautiful home, and that in you and Mrs. Ross I found the most delightful people I have ever met.

Beleive me sincere, you will realize it later on, as God permits you to inspire the hearts and minds, yea the souls of young folks to higher and nobler things. . . .

I am admiringly yours.

G. W. Carver

Carver expressed a similar sentiment of hope for the race in a 1930 letter to the editor of the *Amsterdam News* (New York).

December 3, 1930[20]

My dear Mr. Carrington

Thank you so much for copy of the Amsterdam News which reached me yesterday.

I have read your write-up twice and shall read it again several times. There is one criticism that I would like to make, and that is that you have given me entirely too much credit. I will say, however, that it is the best article that has come out recently. It is safe, sane, and inspirational to young people.

You have an excellent ability for writing, in fact an enviable ability, and as a young man, we are very proud of you. I saw that soon as you came into the "Y" in New York. I have thought about you ever since, and the possibilities that are yours. Young people like yourself get the race somewhere, and I have no fear for the race's succcessful outcome with people like yourself at its helm. We want to multiply more of your kind.

I do not want you to feel that I am attempting to pat you on the back, as it were. I am very sincere in what I am saying. I look forward to your progress with great satisfaction. . . .

Very sincerely yours
G. W. Carver, Director

Undoubtedly the most famous white person to befriend Carver was Henry Ford. Carver derived a great deal of satisfaction from the fact that he, slave-born and of an oppressed race, found his company sought by one of the wealthiest and most powerful men in America. Their first meeting came in

1937, when Carver attended a chemurgic conference sponsored by Ford in Dearborn, Michigan. Carver's reputation as a scientist had, of course, preceded him, and Ford came to visit him at the Dearborn Inn. Their mutual admiration was evident from the beginning. Carver relished writing to his friends about his relationship with Ford. Following the conference, Ford wrote to Carver, praising the latter's work. Carver, in turn, wrote to his assistant, Austin Curtis, Jr.: "I have before me one of the most delightful letters that I have ever read from Mr. Henry Ford. It is a letter that many would give a thousand dollars to get. He says in his closing sentence 'I hope to have the opportunity of visiting with you sometime this winter.'"[21]

A few weeks later, Ford sent Carver a New Year's greeting. Carver responded with unrestrained praise of Ford and his work.

January 10, 1938[22]

My beloved friend, Mr. Ford:

This is just to extend to you greetings, and to say that your personal greetings to me on New Year's day is far more precious to me than a very fine diamond I received. Mr. Ford, believe me sincere when I say that I consider you the greatest man I have ever met.

a. Not alone because you have amassed such a huge fortune in dollars and cents, but in so doing you have brought comfort and happiness to the whole world.

b. The Master Mind coupled up with such a great soul, destined to continue doing marvelous things, because you are in league with the Great Creator of all things.

c. I feel safe in saying you have the greatest exhibition of group and visual education, which is going to sweep the country as rapidly as educators can see and catch your vision.

A happy and prosperous New Year for you is decreed, because you are both the warp and woof of creative forces and every knock to Henry Ford means a boost.

You inspire me every day as I go about my little job, for which I am more thankful than I have words to express.

I think I will have something to show in the utilization of a new waste product here in the South that will interest you, when you come down.

Very sincerely yours,

G. W. Carver, Director

Ford did visit Carver at Tuskegee early in 1938. Carver could hardly contain his enthusiasm over the visit, as is evidenced by the following paragraphs contained in a letter he wrote old friend Harry Abbott on 14 March 1938.[23]

My great friend, Mr. Abbott:

Thank you so much for your fine letter which reached me the other day. It found me very poor indeed from my strenuous trip to Mississippi, but the visit from Mr. Henry Ford on Thursday cheered me up greatly, and I really feel better this morning. I hope that it will last throughout the day. . . .

How I wish that you could have been here to meet Mr. Ford. It was a marvelous event in the history of Tuskegee Institute. He did not hesitate to say who He came to see, and he says that he is coming back again next year. . . .

Very sincerely yours,
G. W. Carver

On another occasion, Carver was Ford's guest at the latter's estate in Ways, Georgia. The event was the dedication of a school named for Carver, who obviously relished the attention Ford showered upon him. The following excerpt is from another letter written to Abbott, this one on 25 April 1940:[24]

My beloved friend, Mr. Abbott:

 . . . . I did have a wonderful time at Ways, Georgia, at the dedication of the Carver School. I was with Mr. Ford the entire day. I don't think he left me fifteen minutes during the entire day. He rode beside me in the car, helped me over rough places, wouldn't let me walk anywhere, and kept people away from me. In fact the dedication exercises were very quiet. The Principal himself did not know just the hour the dedicatory exercises would take place, and I didn't either. After having buckwheat cakes and other things equally delicious for breakfast we got into the car and drove around to various points on his 75,000 acre estate. A wonderful place it is, and the school is really a beautiful thing. We drove around until about eleven o'clock and finally came up to the school and he helped me out of the car, took my overcoat and I held onto him going into the auditorium. Pretty soon the children began filing in. President Patterson went down with Mr. Curtis and myself, so he was introduced and said a few words in introduction of me, and I blundered around and said just a few words (as I wasn't ready) but they said it was fairly good. Then they called on Mr. Ford for a speech and this is what he said: "Well

I am thankful for the privilege of doing this", and sat down. That closed the dedicatory exercises. There were about a dozen outside people there and they right from the community. He didn't let it be known.

Then we drove around the entire afternoon, and took me to the train and saw me off, and I hadn't been home more than twenty minutes until I got a telegram making inquiry as to how I stood the trip. I will tell you more about him when I see you. . . .

Most sincerely yours,
G. W. Carver

Late the next year, with Carver's health declining, and his strength waning, Ford had an elevator installed in Dorothy Hall, the dormitory in which the aging scientist lived. A jubilant Carver wrote this letter of thanks to his friend.

September 29, 1941[25]

The greatest of all my inspiring friends, Mr. Henry Ford:

Please do not think me an ingrate because I have not written you sooner, but I wanted something very definite to tell you, and here it is:

This is the third week that I have been using the marvelous elevator you gave and had installed for me.

What it is doing for me cannot be expressed in words, but God alone, Who raised you up, endowed you with prophetic vision and rare powers of execution will show you that it is a *life saver*.

This letter is written with my own hand, the first one completed in nearly two years. I rarely attempted to sign my name, as very often I could not do it, the heart was so bad. The swelling of the feet and ankles has gone almost entirely and I can walk 50% better than when you were here.

The Great Creator will reward you. I cannot.

The greatest gift I have ever received from mortal man is the time I met you the first time at Dearborn.

I was thrilled and inspired as never before. I have been able to do better work, you seem to be every present with me in my investigations.

When the great destructive forces have finished their work, the new generation of which you have often spoken, will rise up and call you not only blessed, but will make a beaten path to the many, many educational doors that no one has ever opened before.

I must see you this spring, as I have so many things I want to discuss with you.

Kindest regards to Mrs. Ford.

I am most sincerely and gratefully yours,

G. W. Carver

Despite Carver's fame and the praise heaped upon him by many whites, including Ford, he continued to be victimized by white racism for the remainder of his life. There was the well-publicized incident in 1939, in which Carver, then nationally famous and an old man, was refused a room in the New Yorker Hotel until prominent whites threatened legal action, causing the hotel manager to reverse his discriminatory position.[26]

One month after this incident, Carver wrote to a white friend, Lyman Ward, suggesting that he held out little hope for a change in race relations while he was still alive: "Sometimes I feel that an open discussion of the race problem amounts to just about as much as the discussion of war, as we go on fighting just the same."[27] Similarly, in an interview after Carver's death, his assistant, Austin W. Curtis, Jr., reported on an incident in which a group of white ministers from Georgia visited Carver "and they started talking about the race problem." As Curtis remembered it, "Dr. Carver, without looking up, said, 'Gentlemen, what you do speaks so loud, I can't hear what you're saying. Good day.' And that ended the interview."[28]

While Carver was deeply committed to helping the black masses, he was also often critical of them. Not uncommonly, he wrote that blacks were too dependent, that they needed "to be free and independent people" and to "catch the vision" and "stop the jazz and ragtime."[29] Such feelings, and such comments, put distance between himself and the people he sought to serve.

But those feelings and comments endeared him to whites and allowed him to do what he had hoped to do from the beginning: demonstrate, by personal example, that African Americans were capable of creative thought and scientific achievement. His solution to the race problem, then, was the training and emergence of a whole host of George Washington Carvers. "My people must get credit," he wrote.[30] Nothing would succeed like success. Still, the intransigence of white racism plagued George Washington Carver all of the days of his life.

NINE

Carver and His Boys

Dear, you know me well enough now to know that I am sincere when I say I love you, you know why I love you, because you are exceptional boys with an exceptional future ahead of you, long after I have passed on you must work.

Geo. W. Carver 21 November 1930

EARLY IN THE 1920s, members of the Commission on Interracial Cooperation and the YMCA, looking for vehicles by which to ease racial tensions in the South, identified George Washington Carver as someone who could bridge the gap between blacks and whites. The small-school professor suddenly found himself being invited to appear before groups of white youths, believed by CIC and YMCA officials to be the most educable and malleable.

Everywhere that Carver went he captivated his young audiences. Often, he would look out over a group before him and search for a face and eyes that he thought bespoke intense sincerity and spirituality. The eyes were critical, as he explained to a group on one occasion: "You can tell the people who acknowledge God and love men by their eyes. I have seen eyes that could take my breath away . . . . Eyes of Christian souls are beautiful. They send out love, joy, peace. Their light stands out within a group. I can always pick them out."[1]

Carver began to develop warm relationships with many of those youngsters whom he had "picked out." Gradually, he came to identify and refer to those young men in whom he saw great promise collectively as "my boys" They became the bachelor Carver's second surrogate family, in addition to his Tuskegee "children," and he showered them with affection, nurturing them through the crises of growing up and counseling them on career decisions. They wrote to him, addressing him as "Dad Carver," "Dear Father," "Daddy," and "My Only Dad" One young man expressed a sentiment that each of Carver's "boys" must have felt:

> After spending three of the greatest days of my life with you, indeed it seemed much like a dream to me that such a man as you really lived. . . . I wish that it was possible for me to tell you in words what you mean to me. From the inspiration that you give to me through your love and confidence, somehow I reach to higher heights than I had ever thought of. . . . [2]

One of the first of the young people to become a Carver admirer and member of his informal family was Jimmie Hardwick, a Virginian who had been well-steeped in the tradition of white superiority. Only Carver's transparent spirituality had allowed Hardwick first to accept the older man and, ultimately, to venerate him. This early letter to Hardwick is one of the most poignant.

10-29-23[3]

My precious friend Mr. Hardwick:

Your wonderful letter has just reached me. It is evening and I am alone for a few minutes, and I always feel that I must answer your letters right away.

This letter reveals another side of your life. You are now in the midst of a great struggle. You are fighting for freedom, you will win, God is on your side. Continue to follow him, wherever he leads follow. Whenever He says stay away from church or anything else for a while, do so. He will however not permit you to stay away from church very long at a time lest the effect would be bad, people would not understand you and think you were growing cold and indifferent, so therefore you will not get His approval.

My friend I love you for what you are and what you hope to be through Christ Jesus. I am by no means as good as you give me credit for being.

There are times when I am surely tried and am compelled to hide away with Jesus for strength to overcome. God alone knows what I have suffered, in trying to do as best I could the job He has given me in trust to do, most of the time I had to work without the sympathy or support of those with whom I associated. Many are the strange paths God led me into. He is and will lead you likewise.

God has so willed it that there were always a few good friends to encourage and strengthen me when the burden seemed greater than I could bear. God gave you to me for courage, strength, and to deepen and indelebly confirm my faith in humanity, And oh how I thank Him for you, you came to me when I needed you most.

I know the fetters that hold you down. It is now, the beginning of a new era for you.

Do not get discouraged when you do not seem to accomplish all you wish, sow the seed. . . .

One of the most revealing things about the letters written by Carver to his boys is the obvious reciprocity of the relationships. That the young men need-ed and treasured Carver's friendship is obvious; but Carver's dependence on these youths is no less clear.

4-5-24[4]

My beloved friend, Mr. Hardwick-

Your letter touches me deeply, how I wish I was worthy through Christ of all the nice things you say about me.

I can never be to you what you are to me. I love you all the more dearly because you belong to another race, and because God is speaking through you and using you to teach to all the world the Fatherhood of God and the brotherhood of man, and how sweet it is to let God purge our souls of all ego and littleness, and give us a little taste of heaven while here on earth. What a privilege to be in league with the great teacher.

Your visit here lingers with me as a sweet "Savor", because as I said before I saw more clearly what God has in store for you.

I studied you so closely when you were here and it made me very, very happy that I had the privilege of knowing you and being able to call you a true friend, in Christ. Divine love will one day conquer the world; you are to figure very prominently in this conquest.

I trust you will pray for me that I may get rid of all my littleness.

I have just as much to fight as you have but of course not the same things.

The "Love of Christ" is sufficient to transform both of us into what He would have us be.

I love you because you are my ideal type of man. And one that I can confide in at all times.

You are right altogether people are too selfish and self centered to let Christ do much for them. I do feel however that God in His own way is working out the problem through such disciples as yourself.

More and more I feel that you are going to enjoy life more fully, because Christ will have a "larger fullness" in your heart.

The books you have chosen please me greatly. You are certainly moving along the right lines now, couple this with constant observation and little by little as God gives you light you will become a new man.

If you wish to box or to go to boxing bouts I think it is all right, bearing in mind two things from Paul. If by the eating of meat etc. and I am all things to all men etc. God meant that we should play and have a real good time and enjoy the things He has so wonderfully created, and through which He speaks to us every moment if we so desire.

I am looking forward to the time with pleasure when I shall see you at Blue Ridge and have the privilege I hope of a walk so that Christ may speak to us both through the wonderful flora and fauna of that section.

I fear however that we will have very little time to be together, as we will both be very busy.

I am sure they will draw more heavily upon my time, that is what I have been informed already. However we must find some time. I can see more progress in you all of the time continue to let Christ lead, you simply work and follow where He leads, and "day by day in every way" the pastures will become greener.

My friend I am very sure your own brother does not love you any better than I do. I did not have to learn to love you. I did this the first time I saw you, before you ever spoke to me.

It was the Christ in you of course. I can feel your love also; it cheers, helps, and strengthens me in many ways in my work. I never begin a new project but what I think of you in connection with it.

May God keep, guid and prosper you in every way, to His glory.

Is the wish of your sincere friend

Geo. W. Carver

"Postcript"[5]

The clock is striking 10 P.M. Just from chapel, but I must write you this.

Rejoice, my friend rejoice, God is doing a wonderful work for you. He is leading you into paths you know not of.

I heard such a wonderful lecture tonight from a Mr. Fields, a Y. worker in Peru.

I thought about my beloved friend Mr. Hardwick who is some time in God's own time to do some marvelous work in His name. Some way

I feel so happy that I know you, and can talk frankly to you as I do and that you do not consider my letters silly, and foolish as they sound to me. God knows I am honest in what I say to you. I will tell you more of the vision that I believe God has given me for you when I see you again.

I am so glad I can have you as my confidential friend. I tell you and write you and talk to you as I do no one else.

Others would misunderstand me, but I think you understand me thoroughly.

Good night. May God continue to bless and keep you.

By the late 1920s, Carver's speaking engagements were taking him all over the South. His busy schedule, and Hardwick's own YMCA work, put the two men temporarily out of touch. But in January 1927, Hardwick reestablished contact. He received this warm report on Carver's travels.

Jan. 30-27[6]

My dear friend Mr. Hardwick :-

What a pleasant surprise it was to me yesterday when I received your fine letter. Of course I have thought of you every day, but never expected to hear from you directly again.

Some one told me that you were in my school as "Y." Secy. I certainly was happy that you were directed by the Great Spirit to go out there.

I have missed your letters so much and your fellowship and it is indeed refreshing to hear from you. I had a wonderful trip through Va. my first talk was made at Petersburg, made two in Richmond one at Stanton, one at Lexington, at Washington and Lee, two in Lynchburg, and an all day talk at Emory and Henry. Emory Va. that is I first met the Sunday school, then the little tots, went to church, dinnered, out riding for an hour, then a volunteer round table talk in which I attempted to answer questions of many, many kinds for two hours and to a large audiance, suppered, then the main lecture and of course talk afterwards to many who were interested. Some followed me to my room and talked until 11 P.M. so you see I had a very, very busy day.

My next trip will be an 8 or 10 days tour through Miss. I am preparing for that now.

I am asking that a picture be sent you. I fear that I do not know just what sort of a message you want as what would interest Iowa students. Possibly you can make this a little clearer to me.

Yes I would like so much to come back and see the dear friends there but the distance is so great and I find travel so hard and tiresome for me that I do not know just when I can come.

Then too, I am obliged to refuse people who want me every day, on an average. How I wish it were possible to respond to at least half of the requests. . . .

I want you to know that my prayers are always with you and for you.

Trusting that God will bless you this year as never before.

Very sincerely yours,

G. W. Carver

Throughout his years of correspondence with "his boys," Carver revealed a deep religious commitment and a sense of oneness with nature. He urged his boys to follow his example.

Mar. 9-28[7]

My dear, dear friend Mr. Hardwick:-

Your letter reached me today and made me very happy. I am happy because God is (with your permission) drawing nearer and nearer to you.

My dear friend, keep your hand in His, Acknowledge Him in all of your ways and He will direct your paths.

No my friend the lovely things you say about me belong to God not me.

O if you could right now step into "God's little Work Shop" and see what He has permitted me to do, and its effect upon the south, you would marvel.

Some days I do not do a thing during the entire day but entertain visitors, Ministers are coming as well as educators, five schools with their pupils have been here this year.

My friend you are always remembered in my petitions to the Great Creator.

I never forget you, God has too big a Job for you to do. God moves in a mysterious way His wonders to perform, I am so glad that you are beginning to see God in Nature, in the things He has created. . . .

How I would love to see you get to the point where you could commune with God, through the things He has created. Your soul longs for it you will never be throughly happy until you do this. God is driving you to it do not resist Him and drive Him from you.

At one time you noticeably lost His spirit I could not feel it. I felt so very sad, by and by it began to come back, thank God. . . .

Just remember this please, you do not know it now, but you will be conscious of it when it comes, you will tell me.

Just keep your hand in that of the Blessed master and walk by His Side always.

I shall pray that God will draw especially nigh unto you in study, traveling and its results, God Can send you Just to the right people.

My beloved friend, God is permitting me to do so many, many wonderful things.

I cannot attempt to tell you all I want to shew you when you come. . . .

My beloved friend I could write and write telling you how marvelously God is using us, I say us because you and the other dear boys are praying for me. O yes, God is answering your prayers. I Can Just see almost those wonderful soul windows of yours, (the eyes) I have never been so conscious of it as I am right now.

Just think of being asked to lecture before a group of Birmingham (Ala.) elite, Sponsored by the Birmingham Chamber, who had handsome invitation cards printed and sent out about 1000 people, I spoke in the Empire Theater one of the Citys' finest.

Upon my return I found three letters, from Waycross Ga. One by the Negro Business Leaguee, one from the Chamber of Commerce, one from the Kiwanis Club and the Lion's Club. All joining together with most urgent requests for a lecture on the possibilities of the south, Aug. 9–10th.

A number of prominent white people motored over to Birmingham, and took me there and back. Some will go to Waycross Ga. they will take me.

My beloved friend, Continue praying for me, they do help me so much, indeed I cannot get along without them.

The thought of your coming thrills me, I can hardly wait for you to get here, I feel that I must see you I want to shew you such a wonderful fulfilment of scripture in the form of an oil for the production of fat and Jeneral rejuvination of the body, it lays on fat in a marvelous way, clears the complexion and makes a tired run down person feel like a new person.

It is a massage oil, I through God have taken it and done wonders with thin, under weight people, people who were tired and rundown, for "Athletes" etc. etc. When you come I want you to know for your self what it will do. . . .

My beloved friend I want you to know that I never in all of my life was prouder and more happy over you than I am this very moment.

May God ever bless, keep and prosper you as never before, I could write on and on but I must stop.

With a heart so full of love and Jinuine admiration for you, whom God is using and will use more and more.

I am so sincerely yours,

G. W. Carver

In the early years of the Depression, before the assistance programs created by the New Deal were available, Carver made an effort to give food, clothing, and even money to the destitute with whom he came in contact. In 1931, Jimmie Hardwick sent Carver a check to help the cause. Carver responded with this letter.

2-19-31[8]

My Own Great Spiritual Boy, Mr. Hardwick:

Your letter is truly wonderful. I have just returned from a hour's trip in the woods collecting. O, the wonders God has permitted me to find, so many new and strange things.

You seemed to accompany me all of the way. I could not keep the tears out of my eyes. I was supremely happy.

How I felt for those dear young fellows you told me about, who were really deceiving themselves.

How can anyone, dear, be happy in the truest and largest sense when they know not God, filled with prejudices, hate etc. etc. If they could only experience the joy of loving humanity, regardless of race or nationality.

Dear, I know just how it grieves you. I meet it almost every day in some form or other.

I feel so sorry for people who can mistreat anyone.

We can only pray and trust that ere the curtain of life falls that they will become conscious of their folly, embrace God, and be happy.

I am praying for you, my dear, dear boy.

The check. God will bless you in many, many ways for such a great heart. I was overcome myself, and I too thank you for more than my words can express.

Dear, I knelt down by the bedside, where God came to us in the little "den" and prayed for light and direction. All day today I have made a survey of the destitute conditions and found that every family, both white and colored had been helped and was receiving enough to keep the wolf from the door, and if they will work they will get along.

I said, O God, this is your money. What shall I do with it that will bring the greatest returns for him?

The urge comes to return it and let you place it where God will direct it. It may help you to go into a field where you, through him, will bring many to him.

O, yes, the "Red Cross" always needs money, but I believe God has another place for it. This exhibition of your great spirit, Dear, has helped me greatly.

I so thoroughly believe that this check is bread cast upon the waters and will return to you many fold.

Let us watch it. I believe through this you are going to experience a new type of happiness. God will show you.

May God ever bless and keep my dear boy.

So sincerely Yours,
G. W. Carver

Carver also depended upon "his boys" to recruit others into the "family." When additions were made, he responded warmly.

6-26-32[9]

My great spiritual boy, Mr. Hardwick-

The copy of letter which I inclose, will shew you why I have been so happy for the last several days. I know that you with me thank God for this new boy.

He spent all day with me last Sunday, he is such a fine type of young fellow.

I hear right along from the dear little boy we met in Hattiesburg, Johnie Rikle and Mr. Lilly, the Cripled boy. Just had a letter from him yesterday.

I think without doubt that he will come and that through God his disability can be helped.

Dear, I had a fine letter from the young man in Richmond Va. connected with the Cod seed Co. He is studying horticulture I beleive. (I haven't his letter here in the den). He says you asked him to write last year.

I beleive he will make us a fine boy.

Dear, God moves in a mysterious way His wonders to perform.

In your Case He is performing wonders. It is marvelous what He is able to do with you who is willing to follow where He leadeth. . . .

Your great spirit has never left me since your return from Miss. Pray with me, dear on God's direction in the Case of Mr. Lilly Jr. If he can be helped, make the urge so strong that *He* will come.

My urge says right along, Come, you can be helped in more ways than one.

Yours with so much love and admiration,
G. W. Carver

Often, Carver became close friends with the families of the boys he called his own. Such was the case with Jimmie Hardwick's family, particularly with his mother. Carver established a correspondence with her that lasted for more than a decade. She visited him several times at Tuskegee, and he often gave her plants and did sketches for her. The bond they shared, of course, was a mutual love for her son, as the two following letters make clear

Nov. 14-31[10]

My esteemed friend Mrs. Hardwick:-

Thank you so very, very much for your fine letter and the clippings, which are quite unusual, I do indeed prize them for my scrap book.

Articles like these makes me feel so very insignifficant, they have given me so much credit that I do not deserve.

Not a day passes that I do not think of the delightful time I spent in your home last spring.

I am so glad you like the little picture I beleive it will grow upon any one who studies it. I certainly did enjoy making it. I wish, however that I "had had" more time to have put the rounding of it out.

Your dear boy is one of my greatest treasures, the dear, handsome fellow is a constant comfort and inspiration to me.

He is an ideal young man to me.

I hope to see him before long.

It is not at all difficult to tell where his sterling qualities come from.

He is Just good all the way through. One of the strongest testimonials of his greatness is his love and devotion to his brother, nothing could be more beautiful, it is devine.

Sometimes Mrs. Hardwick, indeed very often, I wonder why the Lord has blessed me with such friends, the very best in the whole country. I have done nothing to merit such a gift.

Trusting you will have an unusually thankful thanksgiving.

I am very sincerely yours.
G. W. Carver

My esteemed friend Mrs. Hardwick:-

How happy I am to get your splendid letter. I trust the seed will come up and do well for you.

Of course you love flowers, and I am sure you will adore these.

I hope you will not over exert yourself digging in the garden, regardless of how appealing the desire. . . .

Splendid, that you could have your precious boy home for a whole week. I had the pleasure of being with the dear fellow two weeks. He is a perfect prince, born in him, I was not at all well when we started, but he took such excellent care of me that I returned much stronger than when I went away. . . .

He, through Christ is doing the finest pioneer Christian work I have ever known, the dear boy is making history, (quite unconscious to him) which makes it all the more beautiful.

I wish you could have seen and heard him on this trip, wherever he spoke he seemed to grip the people with his Godly inspired wit and wisdom.

Mrs. Hardwick it is not much the dear boy can get from me, it is Just the other way, his great pioneer Christian spirit inspires me to have the utmost faith in humanity, that God is in the Heavens, and all will be made well.

It seems so odd that little picture should appeal to you in this way I want to say that I never in all My life painted a picture, that I enjoyed more in its execution than that one, (as imperfect as it is).

I do not recall having to erase anything and doing it over. The less erasing you have to do the greater will be the transparancy of the picture.

I am Very sincerely yours,

G. W. Carver

Another of Carver's favorites was Dana H. Johnson, whom he first met in 1930. Dana's brother Cecil worked for the Tom Huston Peanut Company during the summer of 1929. At the time, Carver was serving as a consultant to the company. Cecil and Dana, who had both been raised in Columbus, Georgia, were then undergraduate students at Georgia Tech. On 1 January 1930, the Johnson brothers drove the forty-two miles to Tuskegee, at Carver's request, to visit the professor.

Carver recognized Dana Johnson's artistic skills from the beginning. He immediately liked the young man and decided at that first meeting that he would encourage Dana to pursue a career in art. What followed was a warm relationship of visits and letters that lasted until Carver's death in 1943.

In the earliest extant letter written by Carver to Dana Johnson, he commented at length on several sketches sent to him by the young artist.

Aug. 2-30.[12]

My beloved boy Dana:-

Dear, your letter is so refreshing, and as usual full of wit and wisdom.

My dear boy, no apology is neccessary, I knew you were buisy. I knew that you knew that I love both of you dear boys very jenuinely, expect much in the future of you, and that I would hear from each of you.

I think, dear, that I am disappointed more than you, not that I am not happy that you are rounding out your courses there and getting ready for next years work, which I hope and beleive that you are going to get more happiness and make it count for more than any year of your college life.

I wish so much that I could have had you at least a few more times before you left.

I am really very happy over your sketches, they all shew marked ability.

The shadows in No. 1 extend too far around for your ball to float out in air.

Your light seems to come from behind rather than in front which is always more or less unsatisfactory unless it is by special arrangement for a deffinite effect.

No. 2 has wonderful strength and character. Dear, you are going to be fine in expression I beleive you would be just as expressive at the piano, or whatever instrument you choose to play.

Your color schemes are rich and suggestive. I like this one, very much. The little box or cube is very harmonious in its highly contrasting colors.

Study, dear, the law of contrast, I beleive you are giving to excel in this.

No. 3. is all foreground, you have absolutely no middle distance or extreme distance which are so essential to a pleasing picture.

This is what I intended to take up, when you came again, it was my plan to take you outside and shew you those divisions.

Your green foliage is too solid a green, it needs to be lightened up considerably by putting in shadows and high lights.

It has a number of good points about it, bring it with you when you come over and we will fix it.

This picture is worthy of going on card board and finishing, charming idea but rather a hard subject for you as you saw it.

No. 4. is outline work and very good indeed. You have made only a few lines but have made these few lines express your ideas very well indeed.

I feel so proud of your work, and you have hardly begun.

If you run down home during the summer, do come over.

I have your splendid picture before me as I write. I certainly love those pictures of my dear, handsome boys.

Dear, I am especially glad that you are keeping up your exercises, 100% physical and mental is the slogan for you both.

Do as much sketching as you can, trying anything you wish.

With greatfulness, for you two dear boys.

I am sincerely yours.

G. W. Carver

The genuine affection Carver felt for Dana Johnson is obvious in every letter the scientist wrote to his protégé. Young Johnson idolized his mentor and sought his advice on all manner of things, but the relationship was two-sided: a commonality of interests, combined with an almost mystical intimacy, disallowed any generational gap between the two men who were born a half-century apart.

Nov. 21-30[13]

My beloved boy Dana:-

It is impossible for you dear boys to know how much real joy your letter has brought me, indeed I have had nothing on the trip that has brought me more joy.

Dear, I feel that it is just as nice of you boys to write me.

I am having a marvelous trip, and think of you dear, handsome boys (my boys) all the time.

I was disappointed in the results of the foot ball game, The Ga. boys played well but the Quakers just out played them.

Dear, it is the one thing, (the autumn colors) that have made me think of my truly precious boy "Dana", who is gifted in more ways than one, much more than he realizes.

I have longed for you because of your great spirit and artist and musical soul both of which would have been unceasingly thrilled. You are right, I think I have never in all my life seen such a profusion of marvelous greens, and gorgeous yellows so wonderfully intermingled, especially at Tuskegee. . . .

Don't worry, you are going to astonish yourself one of these days, nothing thrills me more than to have you say that you are making some sketches and that you have tried the golden tree. "Old mamma Nature" as you choose to call her is not going to be ashamed of my darling boy's efforts by and by, because you are going to develop so fast and your keen correct vision is going to cause you to interpret and execute with great correctness.

Fine, you could not have purchaced better books for certain fundimental ideas that will develop your own creative ideas. To be sure you like anatomy, and will like biology microscopy, an in fact all things pertaining to nature. Dear, you are going to be exceptional in landscapes.

I am just studying how I can have my precious boy "Dana" next summer. I want you to do some work in, oil, water collor etc. etc.

Fine that you dear boys go to dances and have a good time socially, you need it, and by all means don't let worry weight you down. I have seen so many students, (old young men) who had let the "tons" of worry of which you so aptly speak weight them down and make them look old, act old and feel old.

Such are invariably like yourself and dear "Cecil" have creative minds and find it hard to adjust themselves to the ordinary routine.

I met one wonderful young fellow at John Hopkins Univ. He has a really creative mind, he is a poet, musician and cartoonist. He is discouraged, I want, and he has promised to come to Tuskegee long eneough this summer for me to get him started. I will love to do it for him.

Dear, I think it so splendid that my precious boy "Dana" is now finding himself, and that others are finding you also. My how I love to see you grow. . . .

Glad you are plodding along and making progress. I can hardly wait for Christmas to come, when I will see my dear, handsome boys. I say dear, and handsome advisedly, I mean it absolutely. This trip has not made me love you any more, because that could not be. I feel more proud of you because I see so much of the future for my dear wholesome fellows, and you are both developing in such a satisfactory manner to me. You dear boys please me from every angle. Coming over Christmas, how fine, If you possibly can stay all night, and bring some of your sketches.

Dear, you know me well eneough now to know that I am sincere when I say I love you, you know why I love you, because you are exceptional boys with an exceptional future ahead of you, long after I have passed on you must work.

You are both so strong, so handsome and so wholesome and ambitious that you cannot help but succeed.

I am so happy that you do not object to me calling you my boys and really feeling that you are my boys. . . .

Dear, you and dear "Cecil" look constantly and carefully after your physical well-being, as you know our slogan is 100% mind and body, you can have both just as well as not.

My precious boys make me very happy, because you are prepareing to do the world's work.

With much love,

G. W. Carver

P.S. How I would love to bring you both up here, go to Niagra and do some sketching.

As was the case with Jimmie Hardwick, Carver got to know, visited, and corresponded with the family of Dana and Cecil Johnson. As always, he nurtured and prodded "his boys" to become the very best that they could become, addressing the bulk of his comments to thoughts on art.

Feb. 21-31.[14]

My beloved boy "Dana":-

Your splendid letter reached me just a few moments ago, and it is needless for me to try to tell you how much I enjoyed hearing from my own dear, handsome boys.

Dear, I had to laugh out loud when I found out who your charming and loveable people were.

I so thoroughly agree with you, and incidentally they were equally happy to be guests of such two charming and loveable boys.

I also agree with them in that you can always depend upon dear mothers and fathers, to always be looking out for the comforts of their children even when the child is not aware of it. . . .

Your father wrote me such an interesting letter devoting much of it to you dear boys and your work of which he is very proud. I do not need to tell you again how proud I am you know that already.

You also know that my heart is set on you two dear boys. I doubt, however, if you have an adequate idea what a comfort you dear boys are to me, all of the time.

I do not mind telling you that you too dear fellows are my ideal of young men from every angle.

I love those creative minds, that are destined to make a mark in the world that is worth while.

Dear, your course suits me exactly best the things to help you in the direction of your greatest talents. After you have had the work in charcoal, it will help you so much along the line I am so anxious for you to take up.

My if I can only have my darling boy "Dana" just as often as possible next summer. If you have a long eneough vacation Easter and come home, I wonder if I cannot have you for at least a day and night.

I am so anxious to take you out and do some sketching from nature.

In the black and white, dear, that is about all a picture is, perspective, light and shadow.

I am also certain that dear Cecil is delving into new ideas, I am happy with what he is doing, Nothing pleases me more than to see those spendid minds of my dear, handsome, wholesome boys develop, you are both moving in that direction now.

My darling boys writing is all right I see nothing wrong with it, I am more interested in the ideas expressed than the mechanics of writing, and dear, you can see that you are improving right along.

I am surprised, I did not know that either of you had much time of your own. I wish you both would put it into outdoor exercise, just as far as possible.

Dear, be very careful with that ankle, that you do not weaken the tendons. When you write again let me know how it is. 100% body and mind is our slogan. I wish I had you here or I was there, or somewhere, so I could do that for you I would soon have the soreness out. Indeed I have done it many times and am yet doing it. My jentleman from Tallassee is still gaining weight, he is truly a handsome figure now.

I rejoice that your Father sees now that your talents were not for electricity, what you become will help in the great scheme of your creative mind. Indeed you Father seems rather enthusiastic over what you are doing He said that he was going to try to get over to see me soon, I hope so and that he will bring Mrs. Johnson also. I want them to spend the day, my Amaryllis are in full bloom now, one pot has 14 large lilly like flowers on it now, positively gorgeous.

No apology is neccessary for my darling boys, I know you both are buisy and when one writes it means the other one too.

To have (D. & C.) with me on a collecting trip would be divine.

I was out a little while today and came in with a whole arm load of Miscroscopic treasures.

I am thrilled with the thought that I may have you some day, as especially am I anxious to have you do a spray of beautiful leaves from nature.

Dear, I think you and dear "Cecil" enjoys the spring time quite as much as I do.

Your artist soul is in tune all the time. I fear I shall have to go to Nassau Bahamas. They keep pressing it upon me. The Govt. is behind the movement now.

Just think if my dear handsome boys were out of school and could make the trip, As I go I will do some collecting. And I do not see now how I could do without my precious boys in working them up after I return.

With much love and good wishes and complementing you upon the spendid literary style of your letter.

I am so sincerely yours,
G. W. Carver

P.S. Dear, I am still workin with my southern marbles I hope to dress one Monday, I do not know when I have seen prettier marble, it comes from North Sea.

Carver, of course, possessed the soul of an artist, of which he spoke in the previous letter. In the letter that follows, he writes of how the artist often has to wait to execute his work until the proper mood strikes.

Jan. 14-32[15]

My beloved boy Dana:-

Dear, Yesterday and today I have thought of my dear, handsome boy continuously.

I took an uncontrollable desire to work on my large rose piece.

I got it out, cleaned it up and started work, everything I did was a failure, no character, no soul, I had to put it away and do something else.

Today is Just the opposite, you can see the roses grow with every stroke of the crayon, I have painted a dark red, deep pink, pure white, cream, and a deep pink showing the back of the rose.

How I would love to have you here. I trust my precious boys are well and that your work goes on nicely. . . .

With much love and good wishes to my dear boys.

Sincerely yours,
G. W. Carver

Whenever the Johnson boys had the opportunity, they visited Carver at Tuskegee. Such visits left the elderly scientist overjoyed.

Feb. 1-'32[16]

My dear boys, "Dana & Cecil":-
    This is just to extend to you greeting before I leave and to say that I have never been greeted with a more agreeable surprise than yesterday when you dear boys came in.
    For some reasen, hard to explain, I believe your visit was providential, you looked in every way so manly, handsome, wholesome, intelligent and just the kind of material to fit into the coming demands of the world.
    You are developing physically as well as mentally, Continue to watch your eating sleeping and exercising with care, walk much, go to the gym. etc. etc. Take bacteriology and the other things we discussed anything that will give you greater educational bredth, and strength.
    By all means continue to make a serious study as to how, best, your creative minds can fit into present day needs.
    I feel so proud of my darling boys, your progress is so marked.
    Dear, "Dana" I want you to do as much drawing, painting, nature study and literary work as possible.
    Dear "Cecil", to follow up that laboratory idea, it may develop into just what we want.
    In all thy ways acknowledge Him and he shall direct thy "Paths".
    I thank God for you daily.

With real love and good wishes.
G. W. Carver

Early in 1933, Dana Johnson took a job as a chemist with the Art Crayon Company and American Artists Color Works, Inc., in Brooklyn, New York. It was a job that allowed him to use his training as both an artist and a scientist, a combination that, of course, very much pleased the eclectic Carver. Meanwhile, Dana was continuing to take art courses at Columbia University.

Jan. 17-33.[17]

My dear boy "Dana",
    Your letter makes me very, very happy. First that your Father is improving, Second, that you have such a splendid job along the very lines that will work into your art work, this is indeed fortionate, and a pretty good salary for these times.

I can see the hand of providence in it and indeed there may be quite a future there for my dear, handsome boy.

Third, your art club, Just the thing, how wonderfully everything is working out.

Your little sketch, dear, is very pretty, I do not understand her work as yet they are so undeveloped that I cannot see just what she is driving at. I dare say she will come out all right, follow faithfully her suggestions and then if you wish you can work in your own originality, which I am so anxious that you do not spoil, but, dear go ahead we will take care of that.

I think she means dear, that your middle distance is not prominent eneough.

With just a few little changes in your picture it can be turned into a beautiful thing, we will transform it when you come over. This club is going to be of inestimable value to you.

You know dear that my heart is set on you two dear boys and I always enjoy myself to the uttermost when you are here with me. I trust dear Cecil is better by this time, wish I was there long eneough to massage that cold out of him.

He like you will do his best in whatever he undertakes. How happy your expression about the little card makes me. Your expression is identical with those of experienced artists, so soft and yet so brilliant. And that the picture grows upon you as you continue to look at it. With so much love for my precious boy.

Geo. W. Carver

P.S. Dear, get all you can from any of them, Just as they want you to take it. I note that she shews talent for cast work. If you want me to send anything back you send me just let me know. I am so proud and happy over my precious boy "Dana".

Carver never failed to encourage Dana Johnson in his art work.

Sept. 9-33.[18]

My beloved boy Dana:-

Your fine letter makes me unusually happy I think it quite providential, dear that you should get such a fine job right along the lines in which you are interested and for which you have such decided talent.

Just remember dear, that if you acknowledge Him in all of your ways He will direct your paths.

I am certainly happy for my precious boy Dana. Fine, no doubt you will work into the firm and bring about some revolutionary measures with reference to color.

I do not know of a better place for my dear boy. Although you are so far away that I cannot see you.

I am indeed happy that you kept up your art work, and that you stuck to it. You are now in position to some real worth while work. To be sure get in touch with and get training from as many masters as you can.

Be in your work just what dear "Cecil" will be in his.

Dear, I am now working on my rose piece I find it so interesting, it is developing nicely. I do not get the time to do much work.

So glad your mother could go the Fair in Chicago. I am sure got much out of it.

I do feel so happy and proud of my dear boys.

With so much love and admiration.

G. W. Carver

In the late 1930s, Carver was ill regularly and spent extensive periods of time hospitalized. Handwritten letters ceased to be a possibility; indeed, he was often unable to sign his name to letters dictated by himself to a secretary. Still, he stayed in touch with a number of his boys, including John C. Crighton, whom he had met for the first time in 1923. In 1939, Carver sent the following letter to Crighton.

July 8, 1939[19]

My beloved boy, Mr. Crighton:

Your fine letter reached me yesterday, and has been read many times, and very touchingly as I am sure that you very well know how I feel about my beloved boy "Johnny" who has meant so much in my life. Much more than you can ever possibly know. You have given me inspiration, information and have been altogether a wonderful idealistic man. It is impossible also to tell you my feelings, when you came into the office, you know very well that you have always been my ideal young man, and you certainly have come up to my expectations in every way.

I am so grateful for the privilege of having such a beloved friend.

With so much love and best wishes, I am

So sincerely yours,

G. W. Carver, Director

Likewise, Carver continued to stay in touch with Dana Johnson. In January 1940, already in his mid-seventies, he wrote to apologize for missing Dana's wedding, explaining, if somewhat melodramatically, that his poor health had prevented it.

January 5, 1940[20]

My beloved boy, Mr. Johnson:

This is just to extend you greetings very briefly. I did get your invitation to your wedding, but at the time it was a question whether I would live or not. I have been at deaths door ever since leaving New York. In fact I was ill when I was there but managed to get home, and have been in a very serious condition ever since. I am now having to be under the strongest medication with three injections every week, and allowed to see no company.

During the time of your ceremony I was not expected to live an hour. Now I am allowed to talk or write but just a few moments at a time. I shall write you more fully when my strength returns. I am beginning to feel that it is going to come back.

It is absolutely impossible for you to know what the lives of you dear boys have meant to me, and what they mean yet. I am sure that you know just what I wish for you and Cecil both, and how thoroughly I want you to understand that you are my boys, and shall write you in accordance with what the doctor permits me. He told me very frankly last week that he had no hopes for me until just a few days ago, that I have begun to respond to treatment.

With so much love and best wishes, I am

So sincerely yours,
G. W. Carver

In one of the last letters Carver sent to Dana Johnson, written less than a year before Carver's death, the aged scientist tried to summarize how much the friendship of "his boys" had meant to him. His dominant passion by that time had become the completion of the Carver Museum at Tuskegee.

February 14, 1942[21]

My beloved boy, Dana:

Your fine letter has been here for several days and I have read and re-read it with so much interest, because it expresses the true character of one of the finest boys that it has ever been my privilege to meet, and to have you and your brother as my ideal young men is appreciated more than you will ever know. I am sure you realize, possibly now as never before, as all these years have clinched this inseparable friendship and admiration. Not a day passes that I do not think of my boys and often wonder just what they are doing.

Since my heart has gotten so weak that it puts my hands out of commission for writing personal letters I cannot do it, so this saves you from getting a number of personal letters written with my own hand. If you have any such you had better hold on to them as I fear my hands will not recover sufficiently for me to do any more writing.

It is such an inspiration to me to watch the progress that you and your brother have, and are yet, making, and the future that will doubtless be yours as young aspiring American citizens who must figure into the building up of this great American commonwealth, which I believe is going to effect the world as it has been effected in no time in the past.

I wish very much you could see the museum now since the art exhibit has been installed, that is, the paintings. I would also be happy to see just what progress you have made with your art work. The present war conditions no doubt will give you a remarkable opportunity along the lines of your present activities.

I have been, and am yet, unable to get but a few feet from Dorothy Hall as I do not dare to undertake walking very far. . . . I have been expecting Cecil to be commandeered into the war service to do chemical work. I wish we would hurry and get over this condition and settle down to normalcy again, but whatever changes come about, I am hoping that my two boys, Dana and Cecil, will get remunerative jobs where they can grow and show from every angle their superior brain and constructive line of thought.

I am especially glad that you both have helpmates that are indispensible to your development, and I wish you to thoroughly understand again that I always feel so proud and chesty that I have you two boys as my own personal, intimate friends.

With every good wish and much love, I am

So sincerely and gratefully yours,
Geo. W. Carver

Carver needed the Jimmie Hardwicks and the Dana and Cecil Johnsons and the John Crightons quite as much as they needed him. Perhaps more. To be adored by white youths in a racist America was deeply gratifying; to be loved as both a father and a brother when one had neither children nor siblings was sublime. Carver gave "his boys" much, touching their lives in many ways. But it is obvious that he received no less from them. His life was much richer and fuller for the love and affection he received.

In the three letters that follow, Carver describes to Hardwick just what he meant when he used the designation "my boys."

1-5-31[22]

My beloved Boy, Mr. Hardwick:

Your marvellous letter reached me this evening, and I have read it twice and will continue to read it from time to time.

I rejoice to know that you fully realize now the meaning of the term *my boy*. It is literally so in a way, but spiritually my own son.

God did indeed cause our paths to meet at Blue Ridge. You are the first boy He gave me. Then the others come as He wills it.

My dear boy, it thrills me to note how well you understand it now.

"Sometimes bigger beyond me", to be sure the greater part lies beyond me. I am absolutely nothing except as God speaks through me. Long after I have passed on into the fullness of Joy, my dear little family of boys with their God-chosen colleagues must carry on, just as Paul's dear little Timothy did.

God sent you down here this time. I have been so happy ever since you were here. It may seem odd to you, but I seem to be in better physical condition than before.

I have not been able to do but little walking for sometime. If I attempted to walk much further than my office I would become so tired that I would have to rest. Jan 1st. I walked all afternoon and several hours in the forenoon collecting botanical specimens.

I also spent this afternoon collecting. I have some marvellous things, all microscopic. How the glory, majesty and power of God comes out as I examine each specimen. I believe He will appear in "that still, small voice." God is going to use you marvellously because you are attuned with Him.

My dear boy, God is unusually good to me. I have so much more than you to thank Him for. I have to face so many prejudices and littlenesses that you do not have to contend with.

How my very soul goes out to people who have not found the first principle of true happiness, Divine Love, which must rule the world.

Isn't it fine that God has chosen you and I to make a contribution to this end. . . . Fine if you can come down in April. This, dear, is not my celebration. It is the celebration of the school's fiftieth year.

It may be that I can get away the early part of May if you can go at that time. We are overwhelmed with work now preparing for the big celebration, which is to be the biggest event the school has had in many years.

I too believe God would bless the trip and give us more boys, with which to carry out the great plan in which and for which you were chosen as pioneer. (A great distinction).

My beloved boy, I believe God is going to bless this movement in ways we know not of at the present time.

I must again say that I thank you again and again for you, because I see the will of God in it all. Little by little, as consecrated souls get together God will be in the midst.

I do feel so happy that you are my boy and understand it so well now,

Admiringly yours,
G. W. Carver

P.S. You are going to do as great a work in raising up souls for God as Paul's dark, handsome little Timothy. I say handsome advisedly because a person physically mentally, and spiritually well developed is attractive and cannot help but make others want to be like them. Why should not the religion of Christ make your face to shine?

1-23-31[23]

My beloved boy, Mr. Hardwick:

Thank God for your truly wonderful letter. I am truly happy and have been ever since you were here.

There is an inexpressable happiness such as I have never experienced before from you. First, you are in tune with the reason why I call you *my boy*. Paul saw so much farther in the distance than his dear little Timothy, handsome, sweet little boy that he was to Paul. He was handsome and sweet because Paul saw the extension of God's kingdom through his own son, Timothy as he called him.

Paul knew that he must soon be gathered to his fathers, but dear little Timothy who had caught that great Heavenly Vision above his fellow, would carry on and convert others and so on down the line, even to this day.

I see the same thing in my precious boys. God is going to use them in a very special way.

I can feel your marvellous spirit as never before. You seem to be with me in my daily work for Him. How wonderful it all is, my friend. God moves in a mysterious way his wonders to perform.

God is using you as never before. Your great spirit is being felt now as never before. The responses everywhere you speak show it. Every person who meets you and writes me are enthusiastic over your great spirit. . . .

I have felt your great spirit since you went away as never before. In fact your visit seems to linger as a great spiritual feast which it was.

You would hardly see much if any change with just one massage. I am anxious for you to have them as often as possible. I can understand something of the great spiritual feast you must have had at the Detroit meeting.

It is to be regretted that such incidents such as you describe with reference to the colored delegates are quite common.

God moves in a mysterious way his wonders to perform. I feel that we are making progress slowly but surely.

Glad you saw my dear boys Virgil and Francis. I shall be greatly interested in the young poet. I trust she will send me some of her poetry.

I would love to go to a few of the colleges to which you refer, providing I have the strength and our 50th anniversary does not interfere. How splendid if I could go.

I hope I can meet Mr. Nelson and will be glad to be of whatever service I can to him.

I wish you could fully realize what your trip has meant and is still meaning to me.

That marvellous prayer you made in my little den. It was one of the most impressive I have ever heard. God just seemed to be present with us.

May God ever bless, keep, guide and prosper my great spiritual boy of whom I am so proud and whom I love so dearly and sincerely.

G. W. Carver

1-30-31[24]

My Beloved Boy, Mr. Hardwick:

What a spiritual feast your letter is to me. I am not surprised at all, because I have never in all my acquaintance with you, more correctly speaking ever since God gave you to me at Blue Ridge, have I felt your wonderful spirit so strongly. I have felt so supremely happy ever since you were here.

That marvellous little prayer you made in my little den, God is blessing it, was so far reaching. Its bredth and depth we cannot as yet fathom.

I believe, my dear boy, that I see it and have seen it very clearly. You were the pioneer of the dear young "Timothys", who God chose to blaze out a new trail, one that would bless all mankind. God cannot build up a great spiritual kingdom upon the foundation of hate, prejudice, greed, etc., etc.

You have, like Paul's dear handsome little Timothy, been in training. And your progress has been most satisfactory. You grew just as fast as you lost the finite and took on the infinite.

What a fine summary you make of God's goodness. Yes, He is all that and more. Isn't it a joy to know that you have a real personal God?

I felt all along that by and by you would fully realize what was meant by *my boy*. You will know by and by that it has a still greater meaning, the only road to Divine happiness.

I do thank God for you and feel so happy that you are coming to your own as God wants you.

I am praying especially for God's presence at your retreat. I believe He will be there over you. . . .

Just remember God chose you first to be one of my boys. Whatever comes of it, you through God are the pioneer, the trailblazer as it were.

My dear boy, God has already used you in a very pronounced way, but the best is yet to come as you grow. And you are now in a position to grow as never before. . . .

My dear, dear boy, I am so glad that you can see how God enables me to select these great souls as my boys.

Thank God for the day that I met you at Blue Ridge, (that sacred spot).

May God ever bless and keep you, guiding you in all your work. I did some microscopic work today. You seemed to be near me. The microscope was near the spot in my little den where we met and you made that never to be forgotten prayer. How God revealed himself through those rotten twigs as one of the boys called the sticks. I am praying that in God's own good time and way He will let us walk and talk with him through the microscope.

> With so much love,
> G. W. Carver

Carver's letters to his boys are powerful expressions of emotion. The old professor, misunderstood by so many, found understanding, acceptance, and even love in the company and the correspondence of his young followers. They were Timothy's all; he needed them to be that. It was the only way he could be their Paul. That role gave him strength in times of weakness, security when he felt threatened, satisfaction when he felt unappreciated. He needed his boys, and he needed them to need him. That mutual need, more than anything else, made his life and theirs seem worthwhile.

TEN

Remembering George Washington Carver

An Intimate Portrait

IN 1989, TWO years after the first edition of *George Washington Carver: In His Own Words* appeared in print, I sought to track down and interview a number of people who had known the gifted scientist during his long tenure (1896–1943) at Tuskegee Institute in Alabama. I was especially interested in talking to some of Carver's "boys" (see Chapter 9). The letters Carver wrote to these young men were among the most intense, personal and powerful letters I had ever read. I wanted to talk to one or more of these "boys" in an effort to better understand their connection to Carver, as well as how they felt about their friendship with him from the perspective of more than four decades' separation.[1]

Working with the staff of the George Washington Carver National Monument at Diamond, Missouri, and the Carver Birthplace Association, I obtained a modest amount of funding from the Missouri Humanities Council (MHC) and the Precious Moments Foundation, Inc., in Carthage, Missouri, to cover the costs associated with the use of a studio and two camera operators/technicians from the KOMU television station in Columbia, Missouri. In early April 1989, we were able to bring Dana Johnson, one of Carver's "boys," into the KOMU studio from his home in Long Beach, California. At the time, Johnson was in his early eighties. He had first met Carver in 1930, as a twenty-year old, when he and his brother Cecil drove the forty-two miles from Columbus, Georgia, to Tuskegee for a visit. Cecil had met Carver while working for the Tom Huston Peanut Company in Columbus. Additionally, we were able to interview a second Carver "boy," Dr. John C. Crighton, then a Stephens College professor emeritus of history living in Columbia, Missouri. Crighton met Carver in 1923 when he, too, was twenty years old.

The interviews with Johnson and Crighton were both informative and inspirational. This made me want to interview more people who knew Carver.

With the help of long-time Tuskegee Institute Archivist Daniel Williams, I identified a number of people still living in or near Tuskegee, Alabama, who had either worked with Carver or had been his students. Funds from the MHC and the Precious Moments Foundation covered the costs associated with bringing the two-person camera/technical crew down from Columbia. At the time, I was employed as the Missouri State Archivist, a full-time job that demanded I minimize my time away from the office. So, I flew to Montgomery, Alabama, by way of Atlanta, and rented a car for the 90-minute drive to Tuskegee Institute. There, I met with the KOMU crew for several days of interviews. All of the individuals interviewed, including Johnson and Crighton, are now deceased.

Our first interview was with Daniel Williams. He was the only one of our interviewees who had not known Carver personally. Born in 1932 in Miami, Florida, Williams was only ten years old when Carver died in 1943. Though he never met Carver, Williams emphasized in his interview that he had been heavily influenced by the Carver legacy. He learned, he said, in the sixth grade, a favorite Carver quote: "Start where you are with what you have. Make something of it. Never be satisfied." That quote had been a source of inspiration for him throughout his life.

Daniel Williams went to work at the Tuskegee Institute Library in 1957. He became the University Archivist in 1966. Thus, he came to know Carver through the voluminous records housed at the Tuskegee Institute Archives, where the photographs, documents, and memorabilia associated with Carver and Booker T. Washington make their stories palpably real. Additionally, by the time I met him, Williams had spent decades talking with people who had known Carver. Thus, Williams knew Carver through his work as well as through the memories of others.

Jessie P. Guzman was the second person we interviewed. Because it was a beautiful spring day in the Deep South, we talked with her outside, on the campus lawn. Guzman arrived at Tuskegee Institute in 1923 to work as a research assistant to Monroe Work, longtime editor of the *Negro Yearbook*. She continued to work at Tuskegee long after Carver's death in 1943.

The next person we interviewed was Elva Jackson Howell. She first encountered Carver when he spoke to a group of Virginia State College students, herself included, in 1923. In later years her husband accepted a job on the Tuskegee Institute faculty, and, still later, she joined the same faculty in 1941.

Elaine Freeman Thomas met Carver as a child during the mid-1920s, when her parents moved to Tuskegee after her father accepted a faculty position there. In later years, she became the director of Tuskegee's Carver Museum. We interviewed her on the museum's front lawn.

William L. Dawson ran away from his home in Anniston, Alabama, in 1913 to attend Tuskegee Institute. He was a student of Carver's, both in the academic program and in Carver's famed Bible study class. He also later became Carver's colleague. Edward Pryce and Theophilus Charles (T.C.) Cottrell were former Carver students from the 1930s.

The oldest of the interviewees was ninety-one-year-old Harold Webb, whose advanced age and frailty necessitated our interviewing him in the living room of his home. He had enrolled in Carver's Sunday school class in 1915.

The memories of Carver shared by the people I interviewed stretched over a thirty-year period, from approximately 1913 until Carver's death on January 5, 1943. Taken together, those memories provide an intimate portrait of a complex and fascinating man.

Multiple themes emerged from the recollections of the interviewees. One was agreement on Carver's physical appearance and distinctive voice.

John C. Crighton was a college student little more than one-third Carver's age when the two met in 1923. Crighton described Carver as "a fairly tall man . . . and very strong. . . . [He had] long, strong hands. He was a very handsome man."

Elva Jackson Howell was among a group of students who had gathered in the Virginia State College chapel in 1928 to hear Carver deliver a talk. She had never before heard of him, describing him simply as "a tall, spare man" with a "high voice." Only later did she learn that she had been "in the presence of one of the world's great scientists." Jessie P. Guzman, recalled Carver's physical appearance in similar terms: "He was a tall man, and a slim man, with a dark brown face with very keen features." The fact that Guzman recalled Carver as possessing "a dark brown face" is itself interesting; Carver's dark complexion marked him as different from many of the lighter-skinned African Americans at Tuskegee, including Guzman and the Tuskegee Institute founder, Booker T. Washington.

Edward Pryce came to Tuskegee Institute from California in 1934 because he knew of Carver's work and wanted to study under him. Pryce remembered Carver as "Very tall . . . , slender, stoop shouldered." Pryce, too, noticed Carver's distinctive complexion: "He was dark-skinned." Even more distinctive, at least to Pryce, were Carver's eyes, which he remembered ". . . were deep set, very penetrating . . . very intelligent." As with others who knew Carver, Pryce remarked on his distinctive voice: "[It] was very high pitched, almost effeminate," although, Pryce added, "he was far from effeminate. [He had] a very penetrating voice; you could hear him up and down the halls."

Adding to Carver's distinctive appearance and voice were the clothes he wore. Crighton recalled that "Most of the time he dressed very plainly, in a black summer jacket and a cap . . . rough shoes, dark trousers." Likewise, Jessie Guzman recalled that "He never dressed very well. He was a bit shabby, except that he always had a tie on and had a flower in the lapel of his jacket."

Dana Johnson, a young white student from Columbus, Georgia, remembered that Carver's clothes "were invariably wrinkled. Blue shirt almost always and grayish or very dark trousers, jacket and a tie, his glasses."

Elaine Freeman Thomas remembered that "He did not ever dress in bright colors. It seemed as though he had the same suit on every day . . . . It was a dull gray . . . . A white shirt. Always a little lapel flower." Sometimes, Thomas remembered, the flower would be a dandelion that Carver picked on campus, signaling that he believed in the beauty and worth of a plant that many regarded as a noxious weed. Although Thomas acknowledged that Carver's clothes "were not pressed," she insisted that they were well-fitted and clean. She also remembered that Carver often wore a cap.

Edward Pryce, the Californian who came to Tuskegee to study with Carver in 1934, was totally unimpressed at his first sighting of Carver, a man he initially failed to recognize. "One morning, in going from Millbank Hall [the agricultural building] to the science building, I saw this old man with a long overcoat and a cap on his head and a flower in his lapel . . . walking toward Millbank Hall. I said, 'What's this old guy doing?'" Someone told him it was George Washington Carver, to which he responded in disbelief, "*This* is the guy I came to be with?" Pryce added, "I left Los Angeles to come here as an ag student because I knew about Carver and his reputation. I wanted to be near him. I thought it might rub off on me a little bit, and here was this old guy. I was so disappointed. . . . He cared nothing for his outward appearance."

Pryce overcame his initial shock at Carver's appearance and soon grew to appreciate him as a master teacher, if somewhat unconventional in his methods. Pryce recalled that Carver "would never tell you a thing. He wouldn't answer your question. He was like Socrates, he would ask you another question, and that would lead you to an answer." Carver, according to Pryce, made his students work for answers. He was, Pryce concluded, "teaching me how to observe very closely."

Carver demanded much of his students, and had high expectations of them. William Dawson, who came to Tuskegee as a thirteen-year-old remembered that Carver "had no patience with students who had made no effort [to learn] on their own."

But Dawson and others also remembered Carver as a teacher with a sense of humor, one who enjoyed joking with his students. Sometimes he would ask a question of a student: "Young man, do you know so-and-so?" When the student began to respond with, "Now Dr. Carver, I think . . . ," Carver would cut him off with, "Now, who accused you of being able to think?"

Harold Webb, whose memories of Carver dated to 1915, remembered a favorite Carver joke, about a straight "A" student at Tuskegee whose mother removed the boy from the school. When asked why she had done so, the mother replied, "he knowed more than the teacher. . . . They had him up there six years and they're still trying to make him say 'tater' with a 'p.'"

Edward Pryce recalled Carver chasing students up and down the stairways and through the hallways of Millbank Hall, which housed his office. Feigning anger that the students had interrupted his research with their loud noise, clattering up and down the stairs and through the halls, Carver would chase them with a rolled up newspaper, swatting them while calling them "little rascals," and warning them to stop disturbing him.

Carver not only taught in the classroom; he taught for many years as well a Bible Study class on Sunday evenings. Harold Webb, interviewed at the age of ninety-one, recalled that he met George Washington Carver in 1915, when he was assigned to his Sunday school class. Webb remembered that he remained in Carver's class for two years: "I learned a lot from him," Webb recalled. "He recited all of the parables to us. . . . " "You see, in his class I learned all about Zachariah skinnying down that Sycamore tree, and I knew all about Daniel and the lion's den, the Hebrew children and the fiery furnace. . . . All of that he made quite clear to us. . . . We learned just about all the parables that could be learned from the Bible."

More than seven decades after leaving Carver's Sunday school class, Webb still remembered Edgar Guest's poem, *Equipment*, which he described as Carver's "favorite poem": "Figure it out for yourself, my lad. You've all that the greatest of men have had. Two arms, two hands, two legs, two eyes; and a brain to use, if you would be wise. With this equipment they all began, so start for the trip and say I can."

Tuskegee Institute legend William L. Dawson also attended Carver's Sunday school class, in the years before he became a musical composer, choir director, and Tuskegee professor. Dawson's earliest memory of Carver was of the "Professor, the only one on the campus called by a title." "Everybody else," Dawson remembered, "was 'Mr.'" Dawson also recalled that Carver's Bible class met in Carnegie Hall, which also housed the campus library: "After what we called supper . . . students would go up there on the second floor [of Carnegie Hall]. They would ask him questions from the Bible, and he would answer them, he would interpret them for you." This, Dawson remembered nearly a lifetime later, "was a wonderful experience." So wonderful, Elva Jackson

Howell remembered, "it was said that if you did not get there on time you would not be able to get a seat."

Carver's teaching of a Sunday school class was only one aspect of his deep spirituality. Everyone who came in contact with him commented on this characteristic of his life. It was especially evident in his relationships with the young white men whom he befriended during the 1920s and 1930s. He called these young men his "boys." They corresponded with him and came to visit him at Tuskegee.

John C. Crighton, a native of Richmond, Virginia, was, as mentioned, one of those boys. He first met Carver in the summer of 1923, when he and a group of students from Lynchburg College in Virginia attended a national YMCA conference at Black Mountain, North Carolina. Carver was one of the speakers at the conference and stayed in a cabin close to the one occupied by Crighton and his fellow students.

Crighton was the editor of his college newspaper, an activist in the peace and Social Gospel movements, and an active member of the Disciples of Christ denomination. One of the first things that Crighton noticed about Carver was that the latter stayed up late every night, "Undoubtedly writing letters to countless friends and associates," but that he rose early in the morning "long before the rest of us got out of bed" to take a walk in the Great Smokey Mountain region. Crighton remembered that Carver would always return from his walk "with a small flower in his lapel . . . [a] very, very small flower, but to him . . . it was representative of the total beauty of the natural world." If Carver was asked why he took his walks, Crighton continued, he would respond, "to talk with God and the flowers." Crighton visited Carver at Tuskegee at least twice and corresponded with him for a decade or more. Carver encouraged Crighton and other of his boys to devote themselves to the service of humankind. He regarded himself as one of Carver's "surrogate offspring," explaining his perception of why Carver referred to him as "my boy": "I am sure Dr. Carver felt much as . . . an Old Testament prophet might feel, that he needed someone, a group of people, to carry on his work, to carry his message after he passed away."

One of Crighton's favorite memories of visiting Carver at Tuskegee was of him taking Crighton into his living quarters in Rockefeller Hall, a dormitory, and reading poetry to him. The reading Crighton remembered best was from the great African American writer, James Weldon Johnson. Carver read from Johnson's book, *God's Trombones*, which reproduces the sermons of early black preachers. Carver selected and read "Go Down Death": "And God said, go down Death, go down. Go down to Savannah, Georgia, down in Yamacraw, and find Sister Caroline. She's borne the burden and heat of the day. She's labored long in my vineyard, and she's tired, she's weary. Go down, Death, and bring her to me. And death didn't say a word, but he loosed the reins of his pale white horse, and he clamped spurs to his bloodless sides, and out and down he rode, through Heaven's pearly gates, past suns and moons and stars, on Death rode."

Dana Johnson, too, remembered having visited Carver in his living quarters, in what Johnson referred to as his "den." As Johnson recalled, "sometimes [Carver] would take out the Bible and turn to the Psalms and read." According to Johnson, the Psalm Carver quoted most often was number 121, which he remembered as, "I will look unto the hills from whence cometh my help."

Johnson had a special connection with Carver because of their mutual interest in art. Their conversations and correspondence through the years evidenced Carver's strong interest in art even into his old age. "He wanted very much for me to be an artist," Johnson recalled, "and he wanted me to be a good artist."

Johnson often showed Carver paintings he had done, and Carver critiqued them. "He would immediately say, 'this is not right. The light is not good,' or 'I like the way you did water, Dana. I think you are going to be very good with water.'"

Carver reciprocated by sharing his work with Johnson. "I considered him one of the world's best artists," Johnson recalled. Of a particular still life that Carver showed him, Johnson stated, "I thought it was just the most beautiful painting in the world. The roses especially were wonderful. He was most gifted at painting roses . . . ."

Carver even shared with Johnson a charcoal drawing that he took to be an image of Carver's girlfriend: "He showed us a drawing of someone we think must have been one of the girls that he went with." Perhaps this was a drawing of Sarah Hunt, whom Carver biographer Christina Vella described as someone with whom Carver had a six-year relationship early in the twentieth century.

Like Carver, Johnson came to realize that it was going to be difficult for him to make a living as an artist, especially as the Great Depression deepened during the 1930s. Johnson was aware that his movement away from an art career might trouble his mentor. "I know," Johnson recalled, "that he had such a struggle between deciding upon art or the science and agricultural career that he did follow. He wanted very much to be this artist . . . but he felt that doing what he was going to do at Tuskegee was for the people of the South and the people that needed his help more than the art."

Notwithstanding Carver's disappointment over Johnson's decision to abandon the goal of an artistic career, he encouraged Johnson at every opportunity, always counseling him to do his best, regardless of the circumstances. Carver tried through his many contacts to help Johnson find a job upon the latter's graduation from Georgia Tech. When Johnson "finally did land a little spot in teaching manual training there in Columbus, Georgia, oh, he was so delighted because at least I had something that I could do that would keep the wolf away from the door."

Carver encouraged Johnson to take advanced training in industrial chemistry. That training, combined with Johnson's interest in art and skill as an artist eventually led him to a job and a career as a color chemist. This, in turn, gave him and Carver an entirely new horizon of topics for conversation and correspondence. "Since I was in the color chemistry business, manufacturing artists' materials, he was interested in that. I would tell him of any new pigment that I had encountered. I remember writing to him about Phthalocyanine Blue and Phthalocyanine Green. They were colors that had just come into the market in the 1930s and were very fine indeed because they were permanent. Most blues and greens of that time were fugitive, so he was quite interested in that, but he had not heard of it."

The Carver-Johnson relationship was close and deep, evidencing Carver's great capacity for friendship and his tremendous generosity toward those he cared about. Long-time Carver friend Elaine Freeman Thomas remembered an incident on campus that drives home this point. According to Thomas, one early morning Carver met a Tuskegee Institute employee named Georgia Poole on her way to work. Carver was quite excited and he cried out to Poole, "run, Georgia, run." Unclear as to the reason behind Carver's call to urgency, Poole nonetheless quickened her pace and followed Carver. As it turned out, he had been observing a night-blooming plant begin to flower, and he did not want his co-worker and friend, Georgia Poole, to miss the sight!

Carver's relationship with Elaine Freeman Thomas evidenced another surprising facet of Carver's personality and lifestyle: he loved children. Thomas came to Tuskegee as a six-month-old baby, when her family moved to the campus in the wake of her father's employment there. As a young child, she visited her father in his classroom often. She also visited George Washington Carver in his laboratory, "just to talk." Although a shy child, she felt comfortable with Carver: "Somehow I related to Dr. Carver, and had confidence in him, and was able to talk with him." Thomas followed him around his lab, tagging after him. "I would watch him work," she recalled." "I talked to him. He touched children. He would pat a child on the head or walk through the museum with his hand clasped to a child. He was a very personable person and enjoyed people."

Dan Williams, the Tuskegee Archivist, emphasized this Carver characteristic when he said simply, "[Carver] loved children." According to Williams,

Carver tended to visit families that had children, and he enjoyed playing with children. "He liked doing little things for their children, giving them a plant, or cutting out a little card for their birthday, or that sort of thing." According to Williams, "the more children you had, the closer he was to you."

Yet another intriguing aspect of Carver's life was the seriousness of his conviction, especially by the 1930s, that his administration of peanut oil massages had a positive impact on massage recipients. Theophilus Charles Cottrell, who came to Tuskegee as a student in 1937, had a personal testimonial about the positive impact of peanut oil and peanut oil massages. As a student, he inexplicably began to lose his hair, with large bald spots emerging on his scalp. One day, Carver spotted Cottrell on campus and invited him to submit to a peanut oil massage of his scalp. Following that massage, Carver gave Cottrell a supply of peanut oil so that he could continue to massage his scalp. "I don't know whether that peanut oil did it or not, but the hair came back," noted Cottrell in a 1989 interview. Some fifty years after his first peanut oil massage, Cottrell, then in his seventies, still had a full head of hair.

Cottrell's contemporary, Edward Pryce, also offered a testimonial to Carver's success with peanut oil. Pryce lived in Sage Hall, near Rockefeller Hall, where Carver resided, while he was a Tuskegee student during the mid-1930s. Pryce recalled, "On Sunday mornings there would be a car that would pull in from Columbus, Georgia, and there was a little white boy, crippled with polio. He would go into Carver's apartment. He would be getting peanut oil massages to build up those atrophied muscles. After a while, I saw that kid come out of the building without the crutches." Pryce acknowledged that it was months, perhaps even a year, before he saw the youth "cured," and that the latter still hobbled, but that he no longer used crutches. Pryce added that this was "not an isolated case. There were several others who did the same thing."

Dan Williams was less sanguine in his assessment of Carver's success in treating polio victims. "Many people came [from] far and near to be treated with his peanut oil," Williams remarked. Still, Williams pointed out that there was no record of any healing that actually took place.

Whether the Carver "boy" Dana Johnson believed in the efficacy of peanut oil massages, he harbored no doubt that Carver did. "He believed very strongly in them," Johnson recalled. "He told us of a number of his people that he had treated . . . . He had actually rubbed the peanut oil into their muscles and helped those arms or legs . . . to recover." Johnson added, "He was seriously sure that his peanut oil treatment for after polio recovery were really working." Johnson also reported that Carver gave him and his brother peanut oil massages. "He would like to give us massages, peanut oil massages, once in a while," Johnson recalled. He added, "I was always skinny, and he wanted to put some weight on me." As Johnson remembered it, Carver had a reputation for helping people who wanted to gain weight to do so.

Carver was well known on the Tuskegee campus for his efforts as a conservationist. Elaine Freeman Thomas remembered that one of her earliest memories of Carver was of him giving her a jar of his peanut "vanishing cream." Carver told her that if she used the cream every night, "you will grow up to be a beautiful lady, you will look like a princess."

Carver sought to teach Thomas and others about conservation of resources. One day, she recalled, "he pulled out of his pocket . . . a ball of twine, and he said . . . 'little girl, this is what you throw away, and many people like you, and you get a package out of the post office, unwrap it, and throw the cord away . . . . I find a second use for many things that you throw away.'"

Among the products that Carver crafted from cord used to bind packages together were macramé creations that adorned his living and working spaces. According to Thomas, Carver made his own neckties out of discarded cloth. He also made a lot of his laboratory equipment from materials retrieved from the Tuskegee Institute trash pile.

Dana Johnson asserted in a 1989 interview, "If George Washington Carver was alive today, he would be considered an environmentalist." John Crighton, felt the same way. In response to the question, "Was Dr. Carver an environmentalist," he responded that he most certainly was, emphasizing that when Carver arrived at Tuskegee Institute in 1896, "His basic problem was to restore the soil of the South to productivity . . . ." Carver had, Crighton emphasized, "the basic philosophy of the environmentalists, of the people of the Wilderness Society and the Sierra Club, in that he had an appreciation of the mystery and the beauty of the universe." Crighton recalled Carver quoting to him "the opening verse of the 19th Psalm that 'the Heavens declare the glory of God and the earth shows forth His handiwork.'"

Viewed in this way, the Earth, to Carver, was "a place of beauty and of mystery, it was God's handiwork, to some extent God's incarnation, you might say." According to Crighton, "[Carver] would today be concerned with problems that did not exist in his time: the greenhouse effect, the erosion or destruction of the ozone shield, and the ravaging of the rain forests, the pollution of our oceans and inland waterways and more."

Those who knew Carver well over a long period of time described him as driven by, even obsessed with, his work. Dana Johnson certainly felt that Carver was "very strongly driven by a sense of mission," a desire to help "his people in the South."

Carver's former student Edward Pryce recalled, "You could see he was purposeful . . . . He was in to what he was doing." Pryce added, "The man was working day and night . . . . [H]e walked as fast as he could walk from place to place . . . there wasn't a lot of casual anything about the man."

Jessie Guzman, whose knowledge of Carver and his habits extended back to 1923, remembered that "[Carver's] concentration was on his work . . . . He really was wedded to his work." A difference of opinion about whether Carver "usually" dined with his co-workers emerged from interviews with his contemporaries. Guzman remembered that he did. "He ate with the rest of the faculty." There were about eight people to a table.

Edward Pryce, on the other hand, remembered Carver usually eating alone: "He would eat alone in the cafeteria. There would be a big round table, six or seven feet in diameter. He would be there by himself." Pryce added, "I don't think he encouraged people [to sit with him]. He couldn't stand a lot of frivolous conversation."

Elva Jackson Howell even recalled that Carver would sometimes eat alone in his room, dining on food he cooked over a Bunsen burner. She remembered a time when she was walking down a hallway past an open door to Carver's dormitory living quarters. There was a horrible odor flowing from the room. Carver was cooking an assortment of wild greens. Seemingly oblivious to the foul odor his cooking had produced, Carver invited Howell into his room to partake of his dinner. Not wanting to offend him, Howell agreed to taste the greens. The experience nearly made her gag, the taste was so bad, although Howell felt compelled to tell Carver the greens were delicious.

The fact that Carver may or may not have taken his meals with his coworkers, and his seeming obsession with his work, that likely inhibited close relationships with his co-workers, further evidences the estrangement between him and many of his Tuskegee colleagues through the years.

According to long-time Tuskegee Archivist Dan Williams, the estrangement began as soon as Carver set foot on campus. Williams maintained that Carver was "ostracized" by his co-workers because he "was the first faculty member [at Tuskegee] to graduate from a predominately white institution . . . . Many people resented that. . . . They thought he should have come from Hampton [Institute in Virginia, *alma mater* of Booker T. Washington], or in the mould of Hampton."

Elva Jackson Howell recalled that some liberal arts professors looked down on Carver because he was "a man of the soil." More pointedly, many thought he was arrogant. As Howell expressed it, Carver was known to be "the first one on the faculty to have a master's degree," something, Howell continued, that "some of his detractors, I suppose, would say he did not let people forget it."

Jessie Guzman just thought of Carver as "very unusual." Although she indicated that she first came to this conclusion because Carver seemed to seek out younger co-workers to socialize with, rather than older ones, she thought there were many things unusual about Carver.

One was his many relationships with white people. "White people liked him," Guzman commented. "There wasn't the usual animosity I think against him as a person as there was against the race as a group." Carver's relationship with whites, especially those who were rich and famous, was a source of pride and comment among Carver's co-workers. Jessie Guzman remembered a visit to the campus by President Franklin D. Roosevelt in 1938. "The main thing that I remember was that the president came in with his attendants and everybody in the community came on the campus to see him and he [President Roosevelt] particularly came I understand at that time to see Dr. Carver."

Similarly, multiple visits to the campus by Henry Ford were also noted by Carver's colleagues. Elva Howell recalled that "Henry Ford came several times because he wanted to hire Dr. Carver."

But the attention and preferential treatment that Carver received, especially from whites, resulted in resentment and jealousy of him by many of his co-workers. Long-time Tuskegee archivist Dan Williams acknowledged "it could be that some of them [Carver's co-workers] were a little jealous—I heard this." William Dawson, who knew Carver for three decades and visited him often in his declining years, was more emphatic: "You'd hear 'em talk, and these were professors. 'The sooner we forget Carver the better we'll be.'" Added Dawson, "They were jealous, that's all it was," [because] "Everyone wanted to see Carver."

And all of those who knew Carver remembered his fervent desire that he and his work be remembered. To that end, he worried over how his work would be continued after his passing and tried to shape the memory of himself and his contributions to humankind. Dawson recalled that Carver "was concerned about his work being carried on." According to Dawson, Carver often recited to him a quote from the Book of Proverbs, Chapter 29, Verse 18: "Where there is no vision, the people perish."

Dana Johnson recalled that Carver had a profound sense of his importance to history. "He had this feeling, right from the start, that he was a man of consequence and he was influencing the Southern farmers and the other manufacturers that were related to him . . . ." According to Johnson, Carver told him and his brother many times, "'save my letters and pictures and these papers and things that I give you because some day you will be glad that you did.'" Johnson concluded, "even then, he knew that he was going to be recognized as an eminent person."

Arguably, one of the most meaningful testimonials to Carver's "greatness" may be the love, respect, and gratitude with which those who knew him spoke of him nearly a half century after his death. John Crighton fought back tears as he expressed gratitude to Carver for helping him "to develop my talents to the utmost." He added, "He encouraged me by example to use those talents in the public service . . . . He flattered me to think that I had the ability to go high if I wanted to." Edward Pryce, also tearful, saw Carver as a role model. "He taught me how to use my brain," Pryce recalled, and he taught me about "dedication." Like so many others whose lives were touched by Carver, they saw their lives as being better for having known him.

As for me, I came away from these interviews with a deeper understanding of and appreciation for the complexity of George Washington Carver. I finally felt as though I had come to know him.

During my short stay at Tuskegee, I roomed in Dorothy Hall, the dormitory building in which Carver lived for several years prior to his final illness. The evening before leaving Tuskegee, I sat on a second-floor porch overlooking much of the campus. With Carver's thoughts from his writings echoing in my mind, and conscious of the words of his friends and colleagues, I penned this reflection in my journal:

> April 11, 1989. Sitting in the sun on a screened-in porch in Dorothy Hall, the building where George Washington Carver lived the latter years of his life. The porch faces West. It is approximately 15' wide and perhaps 6' deep. The porch is just around a corner from Carver's room and the elevator that Henry Ford had installed for him to help him move between his room and his laboratory. Surely Carver must have sat on this same porch many, many times, watching the happenings of Tuskegee unfold before him. To the left of Dorothy Hall, and around a corner, just out of sight, is the Carver Museum, dedicated in 1941. The museum served as Carver's laboratory and office during the years just before his death.

> It is nearly overwhelming for me to be this close to history, to the presence of Booker T. Washington and George Washington Carver. The buildings which make up the National Park here are a wonderful argument for the necessity of the role of the built environment in imparting a sense of place. From this position in Dorothy Hall I can see west, across a

triangle of streets toward a dormitory. To the right, just out of sight, is the chapel, a modern building that looks out of place. It replaced an earlier structure that was destroyed by fire in 1957. Carver must have sat on this porch and watched students going to and from the chapel.

It is still awe inspiring to contemplate the richness of the history we are recording. Many of the people we should have interviewed are gone now, having taken their priceless stories to their graves with them. I am thankful that I was able to talk with these ten; in another decade, they, too, will likely all be gone. Now, at least, George Washington Carver will live on, through their words, and through his own.[2]

Notes

Editorial Policy

1. Carver to M. L. Ross, 7 April 1930, Tuskegee Institute Archives (TIA), George Washington Carver (GWC) Papers, reel 12, frame 0057.

2. Carver to Dana Johnson, 21 February 1931, cat. no. 1705, George Washington Carver National Monument (GWCNM).

Chapter 1/Introduction

1. New York: Doubleday, Doran, 1943.

2. Carver to Mrs. Guy Holt, 23 July 1940, cat. no. 139, GWCNM.

3. Louis R. Harlan, *Booker T. Washington: The Making of a Black Leader, 1865–1901* (New York: Oxford University Press, 1972), 276–77.

4. *Journal of Southern History* 42 (November 1976).

5. William R. Carroll and Merle E. Muhrer, "The Scientific Contributions of George Washington Carver," 1962. This study was sponsored jointly by the United States Department of Interior, National Park Service, and the University of Missouri. Both Carroll and Muhrer were professors in the Department of Agricultural Chemistry at the University of Missouri.

6. New York: Oxford University Press, 1981.

7. Peter Duncan Burchard, *Carver: A Great Soul* (Fairfax, CA: Serpent Wise, 1998); Burchard, *George Washington Carver: For His Time and Ours.* "Special History Study: Natural History Related to George Washington Carver National Monument, Diamond, Missouri." George Washington Carver National Monument, National Park Service, United States Department of the Interior. 2005.

8. Gary R. Kremer, *George Washington Carver: A Biography* (Santa Barbara: Greenwood Biographies, an imprint of ABC-CLIO, 2011), and Mark D. Hersey, *My Work Is That of Conservation: An Environmental Biography of George Washington Carver* (Athens and London: University of Georgia Press, Environmental History of the South Series, ed. by Paul S. Sutter, 2011).

9. Christina Vella, *George Washington Carver: A Life* (Baton Rouge, LA: Louisiana State University Press, Southern Biography Series, 2015); Thomas E. Redy review, *Journal of Southern History* 82 (November 2016): 953–54.

10. The best narrative of Carver's early life remains an unpublished National Park Service report, Robert P. Fuller and Merrill J. Mattes, "The Early Life of George Washington Carver," 26 November 1957, GWCNM.

11. An oral tradition in the family of Oneita Brownfield Byrd Minghini claims that Mary Carver, George's mother, emerged in the Randolph County town of Renick, Missouri, after the Civil War. The tradition further states that this Mary Carver worked for years as a domestic servant in the home of Mrs. Minghini's grandparents and that she spent the remainder of her life looking for her son George. While intriguing, this story cannot be corroborated. The sole evidence for this story is a late-life interview with Mrs. Minghini, a transcript of which can be found in the archives of the GWCNM.

12. For more on Dana Johnson and this interview, see Chapter 10

13. Carver to Mrs. Guy Holt, 13 October 1941, cat. no. 142, GWCNM.

14. For more on Jessie Guzman and this interview, see Chapter 10.

15. Carver biographer Linda McMurry wrote, "Rumors of his homosexuality persist but are undocumented." McMurry, *George Washington Carver*, 245. More recently, Carver biographer Vella states that "something untoward" happened between Carver and Jimmie Hardwick, but offers no conclusive proof. Vella, *George Washington Carver: A Life*, 218. Notwithstanding the lack of corroborating evidence, George Washington Carver has emerged as a subject of a biography in the GBLTQ Encyclopedia (http://www.glbtqarchive.com/ssh/carver_gw_S.pdf). He is also a staple feature of LGBT month (http://lgbthistorymonth.com/george-washington-carver?tab=biography).

16. Evelyn Keller has written extensively about male and female approaches to science. See, in particular, her book *A Feeling for the Organism: The Life and Work of Barbara McClintock* (San Francisco: W. H. Freeman, 1983). One might argue that Carver's "feminine" approach to the study of the environment can be tied into the argument that ecofeminism and feminine science are less domineering and less environmentally destructive. This would be in keeping with the argument of Carolyn Merchant in her important work, *The Death of Nature: Women, Ecology and the Scientific Revolution* (New York: Harper Collins, 1989). See also Glenn Clark, *The Man Who Talks with Flowers: The Life Story of Dr. George Washington Carver* (St. Paul, MN: Macalester Publishing Co., 1939).

17. For more on Elva Jackson Howell and this interview, see Chapter 10.

18. The structure that housed the school attended by Carver remains standing and has been restored. In 2016 it was nominated for inclusion in the National Register of Historic Places: https://dnr.mo.gov/shpo/docs/moachp/Neosho%20Colored%20School.pdf.

19. Fuller and Mattes, "The Early Life of George Washington Carver." More than a third of this manuscript is about Carver's years in Kansas.

20. The pocketknife story is one that Carver enjoyed telling over and over. It is recounted in Ethel Edwards, *Carver of Tuskegee* (n.p., 1976), 9–10. For Carver's version of the story, as reported by Dr. Glenn Clark, see Chapter 2.

21. Carver to L. H. Pammel, 6 July 1923, TIA, GWC Papers, reel 7, frame 0556.

22. This "mighty vision," as Carver described it, allegedly occurred in Washington's first–floor, Porter Hall, office, although some Carver scholars, including Mark Hersey,

are skeptical that it actually happened. The source for this story is Edwards, *Carver of Tuskegee*, 36–39.

23. Ibid., 8; Holt, *George Washington Carver*, 23.

24. McMurry, *George Washington Carver*, 53.

25. George R. Bridgeforth to Carver, 18 January 1904, TIA, GWC Papers, reel 2, frame 0640.

26. Quoted in McMurry, *George Washington Carver*, 69.

27. Interview with Irving Menafee, 16 May 1974, Alabama Center for Higher Education Statewide Oral History Project (hereafter referred to as OHP), 8:8, 41–42, TIA.

28. Interview with Hattie West Kelley, OHP, 7:14.

29. Interview with Elizabeth Ray Benson, 3 July 1974, OHP, 6:7–8.

30. Interview with Jessie Guzman, 26 August 1974, OHP, 5:29.

31. McMurry, *George Washington Carver*, 174.

32. Ibid., 176–78; Holt, *George Washington Carver*, 260–61.

33. McMurry, *George Washington Carver*, 297; Holt, *George Washington Carver*, 259.

34. Carver to Mr. Rarig, 26 July 1932, GWCNM, cat. no. 48.

35. Carver to Dr. Ross, 7 February 1933, TIA, GWC Papers, reel 14, frame 0213.

36. McMurry, *George Washington Carver*, 279–80; Holt, *George Washington Carver*, 309.

37. Abbott to Robert Fuller, 12 October 1964, GWCNM.

38. Carver to Abbott, 29 September 1937, GWCNM, cat. no. 389.

39. Carver to Abbott, 28 January 1938, GWCNM, cat. no. 395.

40. Carver to Abbott, 30 March 1938, GWCNM, cat. no. 398.

41. Carver to Abbott, 2 April 1938, GWCNM, cat. no. 399.

42. Carver to Abbott, 28 April 1938, GWCNM, cat. no. 401.

43. E. N. Hooker to Abbott, 19 May 1938, GWCNM, cat. no. 402; Carver to Abbott, 28 December 1938, GWCNM, cat. no. 419.

44. Carver to Abbott, 26 August 1937, GWCNM , cat. no. 386.

45. Carver to Abbott, 23 August 1938, GWCNM, cat. no. 408. Abbott was not the only person who incurred Carver's wrath when they did not answer a letter quickly enough. On 27 September 1921, Mrs. Milholland received the following two-line note from him: "This is third inquiry written you, why cannot I hear from you. I want to know where you are and how you are getting along." GWCNM, cat. no. 1460.

46. Carver to Abbott, 10 December 1938, GWCNM, cat. no. 411.

47. Carver to Abbott, 27 October 1938, GWCNM, ca t. no. 413.

48. Carver to Abbott, 15 February 1940, GWCNM, cat. no. 440.

49. Carver to Abbott, 27 November 1941, GWCNM, cat. no. 459.

50. This episode is recounted in Edwards, *Carver of Tuskegee*, 119–22. Hardwick wrote a brief foreword to Edwards's book. In it, he asserted that "Ethel Edwards has portrayed the man I knew and loved better than any other writer."

51. This account is drawn from an autobiographical sketch submitted by Johnson when he donated the letters to the Carver Monument, GWCNM.

52. GWCNM, cat. no. 1712; cat. no. 1705.

53. TIA, GWC Papers, reel 61, frame 0238.

54. This often-quoted Carver comment occupies a place of prominence on a wall in the George Washington Carver National Monument at Diamond, Missouri.

Chapter 2/Self-Portraits

1. TIA, GWC Papers, reel 1, frames 0001–6.

2. Ibid., reel 1, frames 0011–14.

3. Although George Washington Carver identified Moses Carver as "German by birth," federal census records for 1860, 1870 and 1880 indicate that he was born in Ohio. The 1880 federal census lists his parents as being born in North Carolina.

4. GWCNM, cat. no. 1430.

5. Ibid., cat. no. 1426.

6. Ibid., cat. no. 1456.

7. Ibid., cat. no. 1450.

8. Ibid., cat. no. 1476.

9. TIA, GWC Papers, reel 35, frame 0179.

10. Ibid., reel 35, frames 0715–16.

11. GWCNM, cat. no. 139.

12. GWCNM, cat. no. 141. Apparently, Mrs. Holt asked Carver to write an account of his years in Kansas. Carver responded with a one-page document that is included as appendix 7b to Fuller and Mattes, "The Early Life of George Washington Carver."

13. Ibid., cat. no. 142. The "orphan child" theme occurs frequently in Carver's correspondence. In 1937 he wrote to another would-be biographer, "One can hardly appreciate what it means to be an orphan child of a race that is considered inferior from every angle." TIA, GWC Papers, reel 11, frame 0482.

14. Rawhead and Bloody Bones were mythical figures of folklore whose exploits of frightening children were commonly told for generations throughout the Missouri Ozarks. Folklorists trace the origins of these tales to 17th century Great Britain. http://americanfolklore.net/folklore/2010/07/raw_head_and_bloody_bones.html.

15. Glenn Clark, "In the Upper Room With Dr. Carver," March 18, 1939, GWCNM, cat. no. 2819-e.

16. TIA, GWC Papers, reel 42, frame 0528.

17. Ibid., reel 43, frame 0206.

18. Ibid., reel 43, frames 0328–29.

19. Ibid., reel 38, frame 0274.

20. Ibid., reel 38, frame 0370.

21. Ibid., reel 40, frame 0551.

22. GWCNM, cat. No. 87.

23. TIA, GWC Papers, reel 39, frame 1069.

24. Ibid., reel 7, frames 1163–64.

25. Interview with P. H. Polk, 8 August 1974, OHP, 7:11.

26. Gary R. Kremer taped interview of Dana Johnson, April 1989. Original in George Washington Carver National Monument; copy in Gary R. Kremer Papers, CA6143, State Historical Society of Missouri, Columbia, Missouri.

Chapter 3/The Pre-Tuskegee Years

1. GWCNM, cat. no. 1389.

2. Ibid., cat. no. 1390.

3. Ibid., cat. no. 103. At the time that Carver met Borthwick, the young wanderer was working as a house servant in the Steely home.

4. Ibid., cat. no. 58.

5. Ibid., cat. no. 59.

6. Ibid., cat. no. 1484.

7. Ibid., cat. no. 1483.

8. Ibid., cat. no. 1480.

9. Ibid., cat. no. 1432.

10. "A Card of Thanks," *Iowa Agricultural College Student*, October 27, 1896, p. 7, Iowa State University Archives.

11. GWCNM, cat. no. 1444.

12. Ibid., cat . no. 1474 .

13. George Washington Carver to L.S. Pammel, April 29, 1918, quoted in Mark D. Hersey, *My Work Is That of Conservation: An Environmental Biography of George Washington Carver*, (Athens and London: University of Georgia Press, 2011), 34.

14. Hersey, 37.

15. Quoted in McMurry, *George Washington Carver*, 40.

16. Original in Iowa State University Archives (ISUA), Copy, TIA, GWC Papers, reel 1, frames 0952–53.

17. Ibid., reel 6, frames 0712–13.

18. Ibid., reel 6, frames 1247–51.

19. Ibid., reel 7, frames 0333–34.

20. Ibid., reel 7, frames 0355–56.

21. Ibid., reel 7, frame 0554.

22. Ibid., reel 10, frames 0308–9.

23. Ibid., reel 7, frame 0845.

24. Ibid., reel 7, frames 1238–39.

25. Ibid., reel 10, frame 0014.

26. GWCNM, cat. no. 484.

27. Ibid., cat. no. 485.

28. TIA, GWC Papers, reel 2, frames 0734–0735.

29. Wallace to Luzanne Boozer, 7 December 1948, quoted in McMurry, *George Washington Carver*, 41.

30. Original in National Archives (NA), Record Group 16, Copy, TIA, GWC Papers, reel 7, frame 0752.

Chapter 4/Tuskegee Institute

1. Original in Library of Congress (LC), Booker T. Washington Collection (BTWC), Copy, TIA, GWC Papers, reel 1, frames 0759–60.

2. TIA, GWC Papers, reel 1, frame 0761.

3. Ibid., reel 1, frame 0762.
4. Ibid., reel 1, frame 0765.
5. Ibid., reel 1, frame 0768.
6. McMurry, *George Washington Carver*, 51.
7. Original in LC, BTWC, Copy, TIA, GWC Papers, reel 1, frames 0771–72.
8. Ibid., reel 1, frames 0831–37.
9. Ibid., reel 2, frames 0192–93.
10. Ibid., reel 2, frames 0435–36.
11. Ibid., reel 1, frame 0788.
12. Ibid., reel 1, frames 0844–0845.
13. Ibid., reel 1, frame 0999.
14. McMurry, *George Washington Carver*, 59.
15. Original in LC, BTWC, Copy, TIA, GWC Papers, reel 2, frames 0930–31.
16. Ibid., reel 2, frames 1028–31.
17. Ibid., reel 2, frames 1058–64.
18. Ibid., reel 5, frames 0087–90.
19. McMurry, *George Washington Carver*, 70.
20. Carver to Mr. Scott, 15 February 1916, TIA, TWC Papers, reel 5, frame 0637.
21. Monroe N. Work, "Review: Biography and Science," *Journal of Negro Education* 13 (Winter 1944): 83–86.
22. GWCNM, cat. no. 473.
23. Ibid., cat. no. 384.
24. Ibid., cat. no. 387.
25. Ibid., cat. no. 395.
26. Ibid., cat. no. 414.
27. Ibid., cat. no. 415.

Chapter 5/The Teacher as Motivator

1. Original in LC, BTWC, Copy, TIA, GWC Papers, reel 4, frames 1080–84.
2. Quoted in Mark Hersey, *My Work Is That of Conservation: An Environmental Biography of George Washington Carver*, (Athens and London: University of Georgia, 2011), 119.
3. TIA, GWC Papers, reel 6, frame 1000.
4. Ibid., reel 7, frame 0443.
5. Original in LC, BTWC, Copy, TIA, GWC Papers, reel 5, frame 0222.
6. TIA, GWC Papers, reel 39, frame 0456.
7. Ibid., reel 35, frame 0744.
8. Ibid., reel 1, frames 0940–42.
9. Ibid., reel 3, frames 1078–82.
10. Hersey, 101–103; for a detailed look at the Nature Study Movement, see Kevin Armitage, *The Nature Study Movement: The Forgotten Popularizer of America's Conservation Ethic* (Lawrence, KS: University of Kansas Press, 2009).
11. Ibid., reel 46, frames 0522–32.

12. Original in LC, BTWC, Copy, TIA, GWC Papers, reel 2, frames 1194–96.

13. TIA, GWC Papers, reel 22, frame 106.

14. Ibid., reel 46, frames 0133–45.

15. Gary R. Kremer and Patrick H. Huber, "Nathaniel C. Bruce, Black Education and the 'Tuskegee of the Midwest,'" *Missouri Historical Review* 86 (October 1991): 37–54.

Chapter 6/The Scientist as Servant

1. GWC Papers, Reel 46, frames 0610–0611.

2. GWC Papers, Reel 46, frame 0619.

3. GWC to BTW, January 26, 1911, reel 4, frames 1021–1023.

4. Original in *Hearings Before the Committee on Ways and Means, House of Representatives on Schedule G, Agricultural Products and Provisions, January 21, 1921. Tariff Information, 1921* (Washington, 1921), 1543–51. Copy, TIA, GWC Papers, reel 46, frames 0889–95. The committee before which Carver appeared was already predisposed toward protectionism, as was much of the Congress in the early 1920s. Prices on virtually all agricultural products had dropped drastically during 1920–1921. In response, Congress passed the Fordney Emergency Tariff Bill, which would have placed high duties on all agricultural imports. Pres. Woodrow Wilson vetoed the bill on 3 March 1921. Subsequently, after Wilson was replaced by the more sympathetic Calvin Coolidge, the Congress met in special session and passed the Emergency Tariff Act of 27 May 1921. The next year the Congress made permanent the temporary tariff hikes of 1921. The tariff did reduce drastically the amount of peanuts imported and caused the domestic price of peanuts to rise slightly. Interestingly, peanut production in the United States was reduced considerably after the duty increase went into effect. Sidney Ratner, *The Tariff in American History* (New York: D. Van Nostrand, 1972), 46–47; Philip G. Wright, *The Tariff on Animal and Vegetable Oils* (New York: Macmillan, 1928), 46–50, 314–15.

5. Moton to Carver, 4 February 1921, TIA, GWC Papers, reel 6, frame 0672.

6. Original in BTWC, LC, Copy TIA, GWC Papers, reel 4, frames 1187–89.

7. Unfortunately, the specific questions asked by McCrarey are not extant. TIA, GWC Papers, reel 16, frames 1239–40.

8. Ibid., reel 10, frames 0679–80.

9. Ibid., reel 12, frames 0033–34. 10. Ibid., reel 13, frames 0228–31.

11. Ibid., reel 13, frames 0601–7.

12. Ibid., reel 12, frames 0041–43.

13. Ibid., reel 19, frame 0947.

14. Christy Borth, *Pioneers of Plenty: The Story of Chemurgy,* (Indianapolis and New York: Bobbs-Merrill, 1939): 226–40.

15. George Washington Carver to T. Byron Cutchin, GWC Papers, Reel 19, frame 0072.

Chapter 7/The Scientist as Mystic

1. TIA, GWC Papers, reel 12, frames 1264–65. Carver's claim in this letter that he and his brother Jim were excluded from church services in Diamond does not seem to be accurate. See McMurry, *George Washington Carver*, 17.

2. Quoted in *New York World*, 19 November 1924.

3. *New York Times*, 20 November 1924.

4. TIA, GWC Papers, reel 8, frames 0444–45.

5. Original in possession of Mrs. Crawford A. Rose, Providence, Louisiana, Copy, TIA; GWC Papers, reel 8, frames 0772–73.

6. Mark D. Hersey, *My Work Is That of Conservation: An Environmental Biography of George Washington Carver*, (Athens, Georgia: University of Georgia Press), 181.

7. TIA, GWC Papers, reel 8, frames 1032–33.

8. Ibid., reel 8, frame 1034.

9. Ibid., reel 20, frames 0819–20.

10. Original in BTWC, LC, Copy, TIA, GWC Papers, reel 3, frame 1227.

11. Quoted in Hersey, *My Work Is That of Conservation*, 185.

12. George Washington Carver, "The Love of Nature," *Guide to Nature*, December 19, 1912.

13. TIA, GWC Papers, reel 10, frames 0635–37.

14. Ibid., reel 34, frame 1097.

15. Ibid., reel 10, frames 0946–47.

16. Original in ISUA, Copy, TIA, GWC Papers, reel 13, frame 0395.

17. TIA, GWC Papers, reel 8, frames 0045–47.

18. Ibid., reel 13, frame 0531.

19. Ibid., reel 35, frame 0066.

20. Ibid., reel 39, frame 0513.

21. Ibid., reel 12, frames 0029–32.

22. This story is recounted in McMurry, *George Washington Carver*, 242–43.

23. There are numerous clippings in the TIA, GWC Papers, reel 61, frames 0238–48.

24. Linda McMurry accurately assessed Carver's "cure" when she wrote: "Undoubtedly the treatment program worked for dozens of people, who were saved from being crippled invalids for life. It worked because it was founded on sound principles of therapy, namely hydrotherapy, exercise, and massage. Such a treatment regime was soon to be made famous by Sister Kinney and became widely accepted. Carver was successful in cases where other people had failed largely because of his skill as a masseur and his ability to inspire hope. His only mistake was attributing his success to peanut oil's power to nourish muscles through absorption. To many this claim raised false hopes of a miracle cure attainable without effort" (*George Washington Carver*, 248–49).

25. GWCNM, cat. no. 491.

26. Ibid., cat. no. 492.

27. Ibid., cat. no. 537.

28. Ibid., cat. no. 535.

29. GWC to Glenn Clark, February 10, 1940, GWCNM, cat. no. 2830.

Chapter 8/Carver: Black Man in White America

1. TIA, GWC Papers, reel 37, frame 0375.

2. Quoted in Fuller and Mattes, "The Early Life of George Washington Carver," 51.

3. Original in BTWC, LC, Copy, TIA, GWC Papers, reel 2, frames 0495–97.

4. GWCNM, cat. no. 1468.

5. Original in BTWC, LC, Copy, TIA, GWC Papers, reel 1, frame 0867.

6. TIA, GWC Papers, reel 10, frame 0652.

7. Ibid., reel 7 frames 0640–41.

8. Ibid., reel 11, frame 0136.

9. Ibid., reel 7, frames 0107–8.

10. Ibid., reel 11, frame 0894.

11. Ibid., reel 10, frame 1066.

12. Ibid., reel 10, frames 1053–54.

13. Ibid., reel 7, frames 0615–16.

14. Ibid., reel 7, frames 0328–30.

15. Ibid., reel 11, frames 0798–99. In 1932, Kester wrote to Carver, "Marvelous are the miracles you have performed in the laboratory but more marvelous still are the miracles you have wrought in the mind and heart of hundreds of men and women." Ibid., reel 13, frame 1010.

16. Ibid, reel 12, frames 0200–03.

17. Ibid., reel 12, frames 0301–3.

18. Ibid., reel 12, frames 0403–6.

19. Ibid., reel 12, frames 0698–701.

20. Ibid., reel 12, frame 0526.

21. Ibid., reel 22, frame 0536.

22. Ibid., reel 23, frame 0406.

23. GWCNM, cat. no. 397.

24. Ibid., cat. no. 447.

25. TIA, GWC Papers, reel 38, frame 0999.

26. For an account of this incident, see McMurry, 275–76.

27. G. W. Carver to Lyman Ward, October 20, 1939, GWC Papers, reel 31, frame 0690.

28. Toby Fishbein interview with Austin W. Curtis, March 3, 1979, Iowa State University Archives.

29. Carver to M. L. Ross, June 28, 1931, TIA, GWC Papers, reel 12, frames 1184–85.

30. Carver to Walter A. Richards, July 1, 1929, ibid., reel 11, frames 1032–33.

Chapter 9/Carver and His Boys

1. "In the Upper Room With Dr. Carver," GWCNM, cat. no. 2819-e.

2. Howard Frazier to Carver, 6 October 1932, TIA, GWC Papers, reel 13, frame 0782.

3. GWCNM, cat. no. 518.

4. Ibid., cat. no. 515.

5. Ibid., cat. no. 516.

6. Ibid., cat. no. 519.

7. Ibid., cat. no. 517.

8. Ibid., cat. no. 569.

9. Ibid., cat. no. 521.

10. Ibid., cat. no. 532.

11. Ibid., cat. no. 520.

12. Ibid., ca t. no. 1728.

13. Ibid., cat. no. 1712.

14. Ibid., cat. no. 1705.

15. Ibid., cat. no. 1729.

16. Ibid., cat. no. 1726.

17. Ibid., cat. no. 1707.

18. Ibid., cat. no. 1703.

19. George Washington Carver to John C. Crighton, July 8, 1939, TIA GWC Papers, reel 30, frame 0002.

20. Ibid., cat. no. 1692.

21. Ibid., cat. no. 1699.

22. TIA, GWC Papers, reel 12, frames 0639–40.

23. Ibid., reel 12, frames 0716–17.

24. Ibid., reel 12, frames 0744–47.

Chapter 10/Remembering George Washington Carver

1. The complete, unedited, versions of these interviews are housed at the George Washington Carver National Monument in Diamond, Missouri.

2. Gary R. Kremer, Journal, April 11, 1989, Gary R. Kremer Papers, CA6143, State Historical Society of Missouri, Columbia, Missouri.

Index